FIRST SEMESTER PHYSICS SURVIVAL GUIDE

The

First Semester Physics Survival Guide

A Lifeline for the Reluctant Physics Student

by Dr. Teman H. Cooke

Three Friendships Company 2013

First Edition

ISBN 978-0-9893205-0-4

Library of Congress Control Number: 2013945197

http://threefriendships.com

Cover: Isaac Newton, the father of first semester physics, sat in an apple orchard one afternoon contemplating the fall of an apple from a tree. He considered what would happen if the apple was thrown horizontally from the tree instead of being allowed to simply fall straight down. From that thought process he realized that the motion of an apple is identical to that of the Moon. From this one brilliant insight we gained an amazing understanding of how things move, both on Earth and in the heavens.

Image by Teman H. Cooke

To my wife Susan, without whom this book would not exist

Table of Contents

Acknowledgments

As much as I'd like to think of this book as my baby, it took a lot of work by a lot of people to bring it into the world. I could not have done it alone. First, and above all, I want to thank my wife Susan. She has been a rock against the shifting sands of self-doubt and fear; without her, I do not think I would have had the strength or courage to complete this journey. It's just one more reason (added to a very long list of reasons) that I love her. Thank you, Susan.

Second, I extend most humble thanks to Charles Cooke and Gayle Wiley, Joseph and Barbara Gascho, Ormond Harriot, Jonathan Hanten, Joon Kim, and the rest of my Kickstarter backers for their generous financial support of the editing and production of the book. Your donations have allowed my work to reach a level of quality and presentation that I would not have been able to attain by myself. Thank you all.

I'd also like to thank the folks at SCORE Lancaster and ASSETS Lancaster for providing training, opportunities, and emotional support as I grappled with how to best produce and market this book. In particular, I'd like to thank Larry Keating and Larry Merris for the sheer amount of confidence and enthusiasm they've shown in both me and my work. Thank you, Larry and Larry[1].

Finally, for general support, comments, and criticisms, I'd like to thank Gail Scott, Jean Kilheffer Hess, Janet Gillespie, Paul Eby-Good, and John Aiken. Your thoughts, suggestions, and ideas have proved invaluable. Thank you all.

1 If there is a Darryl at SCORE Lancaster I haven't yet had the pleasure of meeting him.

"In questions of science, the authority of a thousand is not worth the humble reasoning of a single individual."

– Galileo Galilei

Introduction

If you've picked up this book, chances are you're either currently taking your first semester of physics or you're about to take your first semester of physics. Chances are also good that you're feeling apprehensive about your chances of doing well. If that's the case, this book was written for you.

Regardless of what you may have been told or assumed, physics is not intrinsically a hard subject. Instead, people find it difficult because it requires a detail-oriented, organized thought process. Success, then, is more about learning to think appropriately and less about memorizing equations and example solutions. Once you've mastered how to think about solving physics problems, it doesn't matter what problem is thrown your way – you will be able to solve it.

The jazz trumpeter Clark Terry is well known for his advice to fledgling jazz players: "Imitate. Assimilate. Innovate." It's actually great advice for physics students, too. Toward that end, you will find two levels of discourse presented here. The first encourages imitation. Just as a jazz student learns the fundamentals by copying the masters, you should emulate a physicist to learn the basics of physics. In the following chapters I will provide you with a set of procedures (called **conceptual scaffolds**) that you'll use to solve physics problems. Conceptual scaffolds are designed to hold your hand as you work through the problem-solving process and to help you organize your thoughts. This provides the *how*: the nuts-and-bolts, basic set of tools to achieve physics success. It also gives you a set of steps to copy ruthlessly as you learn the basics of physics problem solving.

However, learning requires more than just imitation. It also requires assimilation – understanding the context and reasoning behind the choices of the steps. For this reason, I've included a second level of discourse to target this assimilation piece. You'll find that I spend a lot of time talking about the *why*: "Why do we use this technique here and not there? Why is this choice preferable for this problem but not that one? Why must I perform the steps in this order instead of the other way around?"

The answers to these questions might not appear immediately obvious when you first read the book, and they will probably not appear immediately useful. However, as you become more comfortable with using and applying the scaffolds these questions should be forefront in your mind. Understanding the answers to these questions is what will allow you to move from simply copying what I do (*how*) to reproducing how I think (*why*). And the only surefire way to get an "A" in physics is to show your professor that you can think like a physicist – that you understand both *how* and *why*.

As for the last part, innovation... well, let's worry about imitation and assimilation before we jump into innovation.

What This Book Is Not

Before I get any further, however, I may need to burst your bubble. This book is designed to help you prepare for your physics course by presenting things in a slightly different way than your professor. It will help you develop and focus your problem solving. Hopefully it will inform your thinking about physics concepts, and maybe even entertain you a little bit with stupid jokes and cheesy lines. However, all of this will only get you so far.

If you're looking for something that will allow you to pass physics without any thought or effort (or, God forbid, get an "A"), then *waves hand*[2] this is not the book you're looking for. If you're hoping to avoid homework, practice problems, and those other annoying time sinks, you should keep looking. If you're hoping to sleep through class without... well, you get the point.

2 Yeah, the old Jedi mind trick doesn't work on my kids, either.

This book isn't going to provide you with a shortcut to learning physics; it will supplement and support you as you do the work you have to do. It will still take a lot of time, brain-sweat, and more mistakes than you will ever be able to count. However, there are three things that can help you get through it all with the least amount of fuss.[3]

First, try to find a way to make it fun. It's physics, not dissection. Speaking of which, how do you know which science lab you're in? If there's something crawling, it's a biology lab. If the room stinks, it's a chemistry lab, and if the experiment didn't work, it must be physics! Ba-dum-bum!

Second, work to actually understand as much as you can. The best way to do this is by jumping in and making mistakes. By analogy, think about a baby who is learning to walk. The kid gets up, wobbles, then falls. It doesn't stop her though; soon enough she's back on her feet and trying again. How many times does the average child fall down before learning to walk reliably[4]? Physics is the same way. Only by standing, falling, and getting up again do you learn to walk; only by learning to walk can you eventually learn to run. Similarly, the only way to become a scientist is to make a bunch of mistakes doing science.

What? You don't want to be a physicist or engineer?

So what?

Seriously! That's the third and final point: Find something about the course that interests you or will pay off in the future. As a doctor, perhaps you'll find it interesting to explore how much force our bones are actually capable of withstanding without breaking[5]. If you want to be a lawyer, physics will give you an opportunity to practice building careful and logical arguments, one step at a time. Or possibly you'll decide that you're actually going to build that trebuchet, automatic can crusher, or rail gun, just

3 Three things besides this book, I mean. Seriously, *waves hand* you need to buy this book.

4 I'm an old man and I *still* can't walk reliably!

5 The answer might surprise you, as might the reason behind it. I'm not going to go into it, but I'll leave you with this teaser: bone is, in many respects, as strong as the reinforced concrete used to build bridges and buildings.

because it's cool. The neat thing is that once you find something that interests you, you'll notice that it hooks into all sorts of ideas and concepts, both inside and outside of the classroom. Instead of being a chore, learning physics will become a new way to unlock your interests.

Conceptual Scaffolds

People building large structures use scaffolds to provide support and stability for the carpenters, masons, and painters involved in the construction. In the same way, conceptual scaffolds provide support and stability for you as you work on building solutions to your physics problems. It's important to understand that conceptual scaffolds aren't solutions, just as physical scaffolds aren't buildings. Instead, the scaffolds support the building process along the way.

Just as physical scaffolds and tools used to build a skyscraper are different than those used to build a barn, different conceptual scaffolds are used for different types of physics problems: Newton's 1st and 2nd Law problems (force problems), conservation of energy problems (energy problems), and conservation of momentum problems. Each section of the book explores one of these in more detail.

All of the conceptual scaffolds are built from the same basic idea, however. They encourage you to use a systematic reductionist approach to solving problems, the same kind of approach a physicist uses. Let's break that phrase – systematic reductionist approach – down to clarify what it means. Because each word generally modifies the one following it, I'm going to start the explanation at the end and work backwards:

- **Approach** – Most people think of approach as getting closer to something. In this situation, think of it as your method of getting closer to a situation. In other words, what actions, skills, and techniques will you use to approach a solution from where you currently are?

  It may seem like a "well, duh" point, but in my experience many physics students are defeated before they even start simply because they can't *approach* a solution. There is a misconception that physics problems are solved with "A-ha!" or "Eurika!" type flashes of brilliance instead of persistence and hard work. Unfortunately, that is almost never the case. In general, your approach to the solution is what determines whether you will succeed or fail, just like an airplane's approach to landing is what determines whether it will land gently or crash. If you never risk crashing, how can you ever learn to land?

- **Reductionist** – The base of this word is "reduce," and that's exactly what we want to do – reduce the problem into smaller pieces that are solvable. If that doesn't make sense to you, consider something you're more familiar with: food. When you are presented with a plate full of food, do you shove the whole thing into your mouth all at once?

  For most people[6] the answer is "No, of course not. I'd choke!" Instead you probably use your eating tools – the fork, knife, and spoon – to break things up into manageable bites, eat each bite separately, and leave behind the things that aren't even edible, like the plate. This is a reductionist approach to eating. In the same way, if you try to tackle everything in a physics problem all at once, you'll metaphorically choke. In either case, it's an unpleasant and completely avoidable situation. Just as you had to learn to use the tools of eating to reduce your food into manageable pieces, you'll need to learn to use the tools of physics to reduce physics problems to manageable pieces as well.

- **Systematic** – Like it sounds, this is about using a system and being detail oriented and organized.

Thus, putting it all together, a systematic reductionist approach is a way to create solutions by breaking the problem down into small, manageable pieces, then dealing with each piece in an organized and sequential way. The conceptual scaffolds are built this way, from the ground up. By using and mastering them, you'll be able eventually to develop your own system-

6 If you answered "Why, yes I do!" please invite me to dinner, as I have *got* to see that.

atic reductionist approach to solving physics (and other) problems. By the way, have you ever wondered why physics is required for pre-med and pre-vet programs? Those are fields that also use a systematic reductionist approach to problem solving. If you learn how to do it in physics, then you'll have a huge advantage over students who haven't learned it.

A Note to Calculus-Based Physics Students

For the most part, I've written this book with algebra-based physics students in mind., but please do not feel that there's nothing here for you. To the contrary, this book can be of great value to you as well. First of all, the only difference between calculus and algebra-based physics courses is how the math is presented – the actual thinking process (and conceptual scaffolds) are the same. However, I've included a few notes specifically to address calculus in physics; you'll find those enclosed in a grey box as shown below.

For Calculus-Based Physics:

This is an example of what calculus-based physics notes look like.

If you're taking an algebra-based physics course, on the other hand, this means you can completely ignore the text in the grey boxes!

Forces

Forces: Newton's First Law

I'm going to assume, at least for the moment, that you've had experiences with cars. If you haven't, go have one now – preferably one that doesn't involve you hitting the car or vice versa, because we're not talking about collisions until the section on conservation of momentum. Instead, find a parked car and watch it.... Don't look away! Now, come back. What did the car do while you were watching it?

If you said something along the lines of "It just sat there," congratulations! You've discovered the first third of Newton's first law:

"An object at rest will remain at rest..."

In other words, stuff doesn't start moving by itself. Hopefully that seems painfully obvious, so let's go back to experiences with cars. This time, I'd like you to do an experiment in your head. (Please do not try this at home!) I want you to imagine you're in a car that's traveling down the highway (at whatever speed seems natural to you). Suddenly, the car hits a big patch of wet ice, just as a deer jumps out into the road in front of the car! The driver panics and slams on the breaks...

What happens? In your imagination, does the car come to an immediate stop?

This kind of situation occurs so frequently that there's a website named (I am not making this up) DeerCrash.org. The very name chosen for the site is incredibly telling. They didn't call it Stopping-Instantly-In-Front-Of--Frightened-Deer.org. This (and personal experience) suggests strongly that the car will *not* come to an immediate stop, but instead continue to move, regardless of what we want it to do. If your mental experiment had you sliding into the deer, you've just discovered another chunk of Newton's first law:

"... and an object in motion will continue..."

But what about going around the deer? Ah, remember the wet ice? If your tires have no traction, how likely is it that your car will be able to turn? Let's add that to what we already have:

"... at the same speed in a straight line..."

Ah! But it is possible to avoid hitting a deer. In fact, it's possible to speed up, slow down, turn corners, slow to a stop in front of a parking space, then start moving again when you realize your car is six inches too big to fit. Obviously, objects don't always remain at rest, and they certainly don't always continue in a straight line at the same speed. So to complete the statement we have to add the equivalent of a huge buuuuut[7]....

"...unless the object is affected by one or more outside forces."

So, putting all of this together, Newton's first law actually says:

An object at rest will remain at rest and

an object in motion will continue that same motion

unless the object is affected by a net external force.

Think about it! If you want to get your car moving, what do you have to do? Push! Or better yet, get the engine to push! Want your brother to stop sitting in your seat? Pull! Exert a force! The only way to get something to move (or once moving, to get it to change), is to push on it or pull on it – in other words, to exert a force.

This is as good a time as any to make this point. Physics, like all subjects, has its own terminology and $5 words. Learning what those words mean

7 I like big "but"s and I cannot lie...

is pretty important, but I'm not going to waste your time with $10 definitions. Instead, here's the kind of definition I prefer:

> **force** (*noun*) – A push or a pull.

If you don't know what a push is or a pull is, just ask any six year old to enthusiastically demonstrate one on you.

Force and Direction

Now, consider the case of the broken-down car. Does it matter where you and your friends push the car? If you have three friends, perhaps it's best to put one on each side, so that they all have plenty of room to push?

If the thought of four people trying to move a car by pushing on separate sides made you laugh, awesome. If not, I'll try harder next time. Seriously, though, it should be fairly obvious that the best way to get the car to move is for everyone to push the same direction.

That's actually a fairly important point. It's obvious, but it bears stating – it really matters what direction you push. The direction of the force is just as important as the strength (or magnitude) of the force.

To remind us that the direction of the force matters, we put a little arrow over the symbol for the force:

$$\vec{F}$$

Mathematically, that indicates that the symbol represents a vector quantity (something that requires both size and direction). For us, it's a reminder that which way you point makes a difference.

External versus Internal Net Forces

Now, if you go back to that statement of Newton's first law above, it is very specific about an external force being required. What does that mean?

Here's another thought experiment for you. Imagine you and three friends are riding in your car when it runs out of gas and stops in the middle of the road. Now, since the only way to get it to the side of the road (and out of traffic) is to push it – would it be better for you and your friends to push the car while sitting inside of it or to get out and push?

If you can propel an un-fuelled car a significant distance[8] forward by pushing from the inside, I would dearly like to see it. (Even Fred Flintstone put his feet out through the bottom of the car to push against the road!) It's generally recognized that the best way to get the car moving is to get out and push. Any push from outside is considered an external force; any push that is completely contained within an object is an internal force. Interestingly enough, it is physically impossible to change the motion of an object with an internal force[9].

By the same token, it is quite possible for two or more forces to cancel each other out. Think back to that example of you and your friends pushing on separate sides of the car. Even if the tires of the car have been greased like a piglet at a pig catchin' competition, it's unlikely that the car will move much. Why? Because you're all working against each other. The person pushing on the front counteracts (and cancels out) the person pushing on the back. So for Newton, it's the total force (a.k.a. **net** force) that matters – how all of your efforts add up or cancel each other out.

---

8 I would consider "a significant distance" to be about the length of the car – 10 to 15 feet.

9 I'll have to add two comments to this. First, right now we are talking about linear motion. When dealing with rotation you'll find that this statement isn't (strictly speaking) true. Second, there are circumstances where it can *appear* as if an internal force causes a change in the linear motion of an object. As we'll see in the section on momentum, however, those changes depend strongly on how you define "internal".

Forces and Contact

I've got a new experiment for you. Set a non-fragile object onto a nearby table, counter, or otherwise clear flat surface. Then, push it across the surface. I don't really care if you go for slow and steady or a quick push that launches away from you. Just give it a push. Exert a force on it to change its motion!

Now that you've done that, return the object to where it started. Now, I want you to move the object again, in exactly the same way... except this time you're not allowed to come into contact with the object. No touching! I'll wait while you work on that.

If you grabbed something else and used it to push the object, congratulations... except you still touched it, albeit indirectly. You lose!

You would be well within your rights at this point if you got a bit cross. "I can't push it without touching it in some way!"

That is exactly the point.

It's actually a rather profound and subtle point, too. Forces require contact to act. If two things aren't touching, then they can't exert a force on one another. For your brother to push you, he's got to touch you. If this weren't true, then separating fighting kids would be useless.

But... what about gravity and magnets? Isn't that a case where an object can pull and push on stuff that isn't touching it? If you check your physics textbook, they even talk about this – "non-contact forces", they call them. This is why I said earlier that it's a profound and subtle point. That non-contact forces stuff is a dodge, used in the same way your parents invoke the Easter Bunny or Santa Claus to avoid getting into specifics. I'm not going to do that to you. Instead, think back to your experiment with the object on the table. When I told you not to touch the object, what did you (probably) do?

Grab something else to make contact for you. Can you imagine if no one could see that item? How awesome would it be to have an invisible stick like this[10]? You could go around pushing things, and to everyone else it would look like you were exerting a force without touching!

Nature has invisible sticks just like this, and they're called <u>fields</u>.

I'm not going to go very deeply into the nature of fields, as they go *way* beyond the scope of this book. However, there are some basics that will help you in your first semester of physics, and you'll definitely run into them again if you have to take a second semester.

First, realize that fields exist. Whenever you see "non-contact force," or "force acting at a distance," remember that between the two objects is that invisible stick – one object pushes or pulls on the stick (field), and the stick (field) then pushes or pulls on the second object. In all cases, contact is required.

Secondly, these fields exist everywhere, although they are limited in what objects they can affect and by how much. The degree to which an object affects or is affected by a field is described by an intrinsic property, the name of which varies depending upon which field you're discussing. For the electric field, for example, it's called charge, and for the gravitational field, it's called mass. I'll discuss the most common forces in more detail in a moment, and if there's a field associated with that force, I'll talk about the limitations of the field (and the intrinsic property associated with it)then.

Third, fields only mediate action between other objects[11]. This means that you'll almost always have an object that's influencing the field to produce some force on the object you're studying. To go back to the analogy, the invisible stick doesn't act by itself. There is always some agent (you, in this case) that acts upon the stick; the stick then acts upon something else. If you ever see a statement like "The Earth acts upon the Moon through the gravitational field," it literally means the Earth affects the field (gravity, in this case), and the field then affects the Moon.

10 It would be a nightmare for parents, however. It would add a whole new dimension to the "STOP TOUCHING ME!" fights that seem to dominate long car trips. "Don't make me pull off my invisible belt!"

11 This is not always true, but it's true enough at this level.

Finally, these fields are physical entities, just like cars, deer, textbooks, people, houses, stairs, planets, dirt... well, you get the idea. All of the concepts we will discuss – forces, energy, and momentum – apply as much to fields as they do to the more mundane stuff we'll talk about. In some ways, fields are even more fundamental than the mundane stuff of springs, boxes, and ramps, but that's a discussion for another day (and another book).

Common Forces in Introductory Physics

Thankfully there are only a handful of forces that you're ever likely to see in an introductory physics class.

Weight

If you've ever stepped on a scale, then you're familiar with the popular idea of weight. Curiously, however, the scale doesn't really measure your weight. (I'll say more about that when I discuss the Normal force.) In physics, weight has a much more specific meaning – it is the force exerted upon you by the gravitational field. Ninety-nine times out of a hundred, the agent influencing the field is the Earth. In fact, you can trust this so much that you should assume every problem describes a situation near the Earth, unless it specifically states otherwise. Don't worry about this too much; generally the problem will say that it takes place "on the Moon," or "near Jupiter," or "in a space station," or the like. That's your clue that you can't make the standard near-Earth assumption.

Gravity is always an attractive force[12] which means that it always pulls. (I'll forgo the usual "Gravity sucks!" joke here. If you haven't heard it, I'm sure you can come up with something on your own.)

Mass is the intrinsic property of stuff that describes how strongly the gravitational field affects that stuff. The higher your mass, the harder a given gravitational field will pull on you. Mass, however, is kind of funny.

12 As far as we currently know, although some physicists argue that there should also be a repulsive (or pushing) piece. As far as I know, however, there's no definitive evidence either way. Stay tuned for more information...

Not "ha-ha" funny, although I'm sure there are exceptions[13], but "that's strange" funny. But, once again, I am going to make you wait until we have more information! (If you can't stand the suspense, skip ahead to the section "What is Mass?")

The weight of an object is always equal to the mass of the object multiplied by the gravitational field in the object's immediate vicinity. Near the Earth the field is effectively constant, mostly because the Earth is very big compared to everything else we deal with on a daily basis. So,

$$\vec{W} = m \cdot \vec{g}$$

where W stands for weight, m is mass, and g is the gravitational field. Remember, the little arrows over the symbols indicate that the direction of each matters; that is, weight always points in the same direction as the gravitational field. Near the Earth, the value of the gravitational field is about 9.801 m/s^2, and points toward the center of the Earth. Just in case you're not familiar, in most cases we call that direction "down." Far away from the Earth, the field's value and direction are more complex.

Normal Force

Think about the surface supporting you – either the seat of the chair you're sitting in, or the floor that you're standing on. If you're anywhere near Earth (which, reading this book, I'm guessing you are), you should be affected by the gravitational field near the Earth. In other words, the only thing that keeps you from falling is that surface below you[14]. We can use this as an example of invoking Newton's first law. I'll walk you through it.

You're not falling. You're probably at rest and remaining at rest, or you're moving at a constant speed in a relatively straight line (if you're one of those people who can walk at read at the same time). Either way, Newton's first law says you aren't experiencing any net external force. But, you

13 For example, "Does light have mass?" – "Of course not; it's not even Catholic!"

14 If, for whatever reason, you are currently in orbit: First, congratulations! Second, you actually are falling, but don't panic! Your sideways motion will keep you from hitting the Earth's surface.

know that your weight (the gravitational force) is pulling downward on you. So, for the total force on you to be zero, there must be another force pushing in the upward direction, canceling gravity's pull! That force is the normal force.

Basically, the normal force results from an object's resistance to being broken or crushed. We could go into great detail about molecular bonding and atomic repulsion, but really the point is that the floor or chair doesn't want you to go through it, so it pushes you out. A scale can't measure how hard the Earth pulls downward on you. It can, however, measure how hard it has to push upward to keep you from crushing it. As long as you're not accelerating (say, by jumping up and down or standing in an elevator) that force should correspond closely with your weight.

Note that a surface doesn't give a damn if you slide across it – that's friction's job (and we'll talk about that next). The normal force is only there to keep you from messing things up. Thus, it always points perpendicularly straight out from the surface. A final point of caution regarding direction: The normal force is not always straight up! If the surface tilts, the normal force will tilt as well!

There is no set equation for the normal force because the surface will only provide as much force as necessary to keep from breaking. In general, only the direction can be immediately determined. You'll have to find the size of the force by solving the problem.

In this production's equations, the normal force will be played by the letter N.

Friction

For the most part, friction gets a bad rap. People focus on the fact that friction leads to heating in your engine, keeps your car from coasting forever, limits efficiency, and prevents mechanical perpetual motion machines. But really, friction is one of the best forces out there. Without friction, you couldn't walk, run, drive, or pick stuff up. Even something as simple as keeping track of your keys would be infinitely harder if friction didn't exist.

So what is friction, then?

The surfaces of all objects have tiny imperfections, no matter how smooth they appear. It's almost like they're covered with miniature mountains and valleys, too small to be seen or felt. However, when two surfaces come into contact, these imperfections catch on each other. Trying to slide them means that the tiny mountains of one surface are continually colliding and bouncing on the tiny mountains of the other, slowing the motion. In effect, this series of tiny collisions give rise to the effect we call friction. This is why often we use a lubricant. The liquid (or something close enough) forms a barrier between the surfaces to separate them, making it much less likely that those surface imperfections can catch on each other.

Physicists talk of two kinds of friction. Kinetic friction acts when the two surfaces are moving against each other and seeks to stop that motion. Static friction, on the other hand, acts to stop the two surfaces from even starting to move against each other. Under otherwise equal circumstances, the force of kinetic friction is usually lower than the force of static friction, mainly because when the surfaces aren't moving those tiny mountains are much more likely to become stuck.

As you can imagine, figuring out how to calculate friction's effect depends on a lot of different factors: the material making up each surface, the overall roughness or smoothness of each surface, the presence or absence of a lubricant, plus all of the properties of the lubricant itself – all of which must be determined experimentally. All of this complexity is reduced down to a single number, μ, called the **coefficient of friction**. Under otherwise identical situations, a smaller μ will result in a smaller friction force. (μ is a Greek letter, and is pronounced "mew," just like the noise a kitten makes[15].) Unless μ is given to you, there is no reason on Earth you should know the value of it when you start a problem. Therefore, you may be given a table of values for different combinations of materials under a variety of situations. In fact, there should be two coefficients for each combination: one for static friction and one for kinetic friction.

15 ...which reminds me of a joke: Which cat will slide off of a sloped roof first? The one with the smallest μ!

One thing friction always depends on is how hard the surfaces are pressing against each other. The harder one surface presses against the other, the closer the two surfaces get, and the more likely their imperfections will catch on each other. We've already talked about the force exerted by a surface that's trying not to be crushed – that's the normal force. So, to calculate the force exerted by friction – f – we need both the properties of the surfaces – μ – as well as the force pressing them together – N:

$$f = \mu \cdot N$$

Now, there are a few things to point out about this equation. First, note that the variables don't have the little arrows over them. That's because the direction of the friction force is never the same as the normal force.

To determine the direction of the friction force, let's consider the source. Remember those tiny mountains colliding? That actually gives us a clue! If we don't try sliding the surfaces, then friction doesn't matter (because the imperfections won't hit each other). A-ha! Friction tries to stop sliding! So the direction of the friction force will always be parallel to the surfaces, trying to keep them from sliding against one another.

The second thing to point out is that the equation above does not distinguish between static friction and kinetic friction. The two basically use the same equation as long as you remember one important point: Friction always stops sliding. It will *never* start a slide. So, if an object isn't sliding, the force of friction (static, in this case) will only exactly cancel out any other forces, if it can. This might be less than the friction force the above equation might give you. For this reason, most physics books put in a less-than-or-equal-to sign instead of an equal sign, but only for static friction. If you understand how friction works, however, you only need the one equation, which will save you some space and hand cramps when you're writing out your equation cheat sheet.

Tension

Tension is that feeling in your shoulders and neck when you think about that upcoming physics test that you haven't studied for. Oops, wrong definition!

Actually, it's not that far off. What you feel in your shoulders and neck are muscles that are prepared for action by being slightly taut and ready to pull. That's the best description of tension – a pull. We don't usually attribute it to muscles, however; most of the time it's from a rope, cord, or some other flexible thing that can't realistically push. Like normal force, there is no set equation for tension. It almost always depends on the situation. However, the direction can usually be found easily: What direction does the rope pull?

Tension's usually denoted by T. At least, it is by normal people. Anyone who tells you differently is one of "them."

Elastic (or Spring) Force

So far, we've talked about objects that don't change shape. When you push on a wooden table, for example, it doesn't visibly deform. Some objects, however, will squish when you push them, and then return to normal when you let them go – for example, a pencil eraser, or your little brother. Similarly, there are some things that you can pull on and stretch, but when you release them they snap back to normal. Examples of this include rubber bands and most people's morals.

The prototypical example of this kind of behavior is a spring, so sometimes elastic force is referred to as **spring force** instead. If you push in on a spring, it'll squish in, but you can feel it pushing back, trying to return to normal. That force pushing back is the elastic force. If you pull on a spring, it'll stretch, but you'll feel the spring fighting the entire way as it tries to pull itself back to normal. Again, that force is the elastic force.

What's interesting about the elastic force is that the size of the force increases the more you stretch or squish the spring. That is, the more you displace the end of the spring, the more the spring will fight to return to normal. This idea is summed up in Hooke's Law, which is the equation form of what I just said:

$$F_{elastic} = -k \cdot (x - x_0)$$

$F_{elastic}$ is the elastic force (I hope). The stuff in the parentheses, $(x - x_0)$, is the distance the spring (or other object) is compressed or elongated; in other words, it measures how much it has been squished or stretched. The k is the **spring constant**. Like the coefficient of friction, it depends on a whole bunch of factors. Like the coefficient of friction, unless someone gives it to you, there is no reason on Mars you should know the value of it. Finally, the negative sign at the beginning of the equation represents that restorative element. When you squish or stretch an elastic object, it wants to restore its original shape – it tries to return to normal. This means that it will always push (or pull) to oppose the squish (or stretch). The negative sign represents that opposition.

Note: Springs can get tricky, depending on how you decide to work with them. There are two basic ways you can handle them. The first is to use the equation only to calculate the amount of force. To do this, ignore the negative sign in the equation, and worry only about the amount of the squish or stretch, not whether it's negative or positive. Determine the direction the elastic force points by considering whether the spring is stretched or squished. Conceptually, this method is the easiest way to handle elastic forces, and works really well if you start out knowing the squish/stretch distance at the beginning (i.e., it's provided in the problem statement).

If you have to figure out that distance as part of the solution, however, DO NOT DO WHAT I JUST DESCRIBED. As one of my favorite physics professors would say, you will "end up in the sticky bushes with no way out." Instead, you will need to solve the problem using the full physics displacement method. In other words, you'll not only have to use your coordinate system to determine the direction of the force[16] but also the direction of the squish. In this case, the negative sign is not only useful, but necessary.

For Calculus-Based Physics:

Springs are often used to model simple harmonic motion – the repetitive back and forth that often characterizes the question-and-answer pattern at a pre-exam

16 See the conceptual scaffold steps 3 and 4 for more information.

study session. This always involves trying to figure out displacement as part of the solution so you won't be able to use the conceptually easy way out. Sorry!

Applied Force

Occasionally, you'll run into a force that doesn't really fit into one of the above categories. These are generally things like "a pushing boy" or "an unknown force." In this case, we have a catch-all category – the applied force. Sometimes this will be referred to as "an external force". The name or source doesn't matter as much as the fact that you recognize the force and give it a name. Otherwise, you can treat these the same way as any other force we've talked about. Again, unless the information is given to you in the problem, there's no reason on Jupiter you should initially know the size or direction of these kinds of forces.

For Calculus-Based Physics:

If you're going to get a time-dependent force that requires a bunch of integration, it's quite likely it'll be an applied force. Remember those elastic forces we were just talking about? Well, apply a force to one end of a spring to make something attached to the other end wiggle back and forth – sorry, oscillate – and you're looking down the barrel of a driven harmonic oscillation problem.

Other Forces

Electric force: The electric force comes from field interactions, just like the gravitational force. In fact, the electric field and the gravitational field are almost identical. The biggest difference is that there is only one type of thing that catches the attention of the gravitational field (mass), but there are two types of things that catch the attention of the electric field. We refer to both of them as charge, but we differentiate between them by calling one type positive charge and the other negative charge.

It may seem to be a bit of an exaggeration to call this two types – after all, I could describe them as simply opposite sides of one type of thing. However, that implies that the one thing for gravity – mass – should also have opposite sides. So where's the negative mass? It's never been observed. Ever. So why should charge span both positive and negative, whereas mass only spans positive? That suggests that perhaps the difference is deeper than just a sign. In addition, the other so-called fundamental fields have even more types that engage them. For example, the strong field, which governs nuclear interactions, interacts through a property called color, of which there are three types – red, blue, and green. (If you've had any experience with color theory or understand how screen pixels make colors, these names shouldn't surprise you.) The relationship between these three types is much more complex than simply plus or minus. Is it such a bad idea, then, to consider positive and negative charges as two different types of the same thing?[17]

In any case, just as gravity pushes mass, the electric field pushes charges. However, unlike mass, we have to take into account the type of charge. There's an easy way to think about this: positive charges are positively compliant. The force they experience is in the direction of the field, and they generally "follow the rules." Positive charges respond to the electric field, just like masses respond to the gravitational field; if you move everything you know about gravity over to electric fields and positive charges, you won't go wrong.

Negative charges, on the other hand, do the exact opposite of those nice positive charges. Negative and contrary, they are considered the bad boys of physics[18].Don't get mad at them, though; that's just how they're made. In addition, it makes them extremely predictable – if you know how a positive charge would respond to a given situation, then you also know how a negative charge would respond.

Electric fields usually aren't tackled until your second semester of physics, along with magnetic fields. This is one of the reasons that physics, part 2, is often referred to as "E and M" or "E-Mag." The "E" refers to study of the electric field; the "M" or "Mag" refers to study of the magnetic field.

---

17 If you're also wondering why charge is both positive and negative but mass is only positive, welcome to the club!

18 They also smoke cigarettes and wear leather jackets with scuffed jeans. Oh, and definitely don't insult one, because it will cut you, man. Seriously.

Magnetic force: The magnetic force is a field force like gravity and the electric force. However, it is much more complicated. I hate to spoil the big surprise from second semester physics, but I'll do it anyway: The magnetic field and electric field are actually two halves of the same field – the electromagnetic field. Because of this, magnetic interactions with charges tend to be much more complex than electric interactions with charges. In any case, I'll leave the in-depth discussion of magnetic fields to its proper place: the sequel[19] to this book.

Thrust: Thrust is the force exerted by a rocket or jet engine on the rest of the vehicle. Usually this vehicle is airborne, although there are situations in which individuals decide to apply this force to a ground vehicle or sea vessel. I'll leave discussions of the wisdom of these decisions to those who have to clean up afterward.

In all cases, thrust involves throwing something out, usually at high velocity. Because of this, rockets and jet planes are better treated using conservation of momentum. However, if "thrust" shows up in a problem with no mention of how much mass was tossed out (the exhaust, ejected mass, or something like that), then more likely than not they expect you to treat thrust using forces. Simply add it as an applied force (see above), and move on.

For Calculus-Based Physics:

Drag: Physics teachers have a love-hate relationship with drag. On the one hand, it has clear, observable consequences that students can imagine, and it makes nice predictions that are backed up by physical experience. In addition, it's one of the only forces that absolutely requires the use of calculus if you want to find any sort of solution. Unfortunately, it absolutely requires the use of calculus if you want to find any sort of solution. Most students have enough problems simply finding the solution to a differential equation that's already given. Drag problems require you to build the differential equation yourself before you can even try to solve it. Generally you'll find professors will demonstrate the process, but not expect you to do it yourself.

19 I am in no way promising or even hinting that I will write that book.

Physically, however, drag is fairly easy to discuss. Remember that friction is due to tiny collisions between the surfaces of two solids. Drag is the result of tiny collisions between the surface of a solid object with the molecules or atoms of a liquid or gas. (Both liquids and gases are called **fluids** because they behave very similarly.) All of these tiny collisions impede an object's motion through the fluid; the object is literally trying to shove the molecules of the fluid out of the way. The faster the object is moving, the more collisions will occur, and the quicker those molecules will have to make to get out of the way.

It's kind of like trying to walk through the ball pit at a fast food restaurant. Even though the balls are much smaller than you, forcing them out of your way slows you down. Trying to run is worse, because by trying to move faster you have to move the balls faster as well. Unlike friction, drag depends on your speed.

Drag gets even more complicated if you consider the type of fluid (is it dense or viscous?), the shape of the object (is it streamlined?), and experimental results (it's the square of the speed that usually matters). Taking all of this into account can easily produce problems that are only solvable numerically – through computer simulation. However, if you are interested in the gritty details of a solvable drag problem, I've included an example in the appendix appropriately titled "Calculus Example." Please see page 229.

If you've read straight through this far, now might be a good time to take a break. Although we've more to discuss for Newton's first law, it's time to stop for a moment and head off in a different direction.

What is Motion?

Okay, you're back? Emptied your bladder, gotten a snack, etcetera? Awesome!

If you recall, Newton's first law not only talked about forces, but about motion as well. I've also been throwing around words referring to motion – resting, moving, speeding, and so on. But really what does it mean that something is moving or at rest? What *is* motion?

If you're like most people, your answer will be something along the lines of "motion is going from one place to another." But there are a lot of things implicit in that common sense answer that we need to unpack.

First of all, motion implies a change of location – "one place to another". Let's start there. Imagine you're staying at a hotel, in room 301. Your friends are at the other end of hall in room 399. To get from your room to theirs, you'd move down the hallway. But consider – is it possible for you to get to their room without, at some point, being in front of room 350? What about room 322, or 395, or room 315? Short of going through a window or taking the stairs, you'll have to pass all of those. In fact, to get to 399, you'd start at 301. You'd then pass 302, then 303, then 304... Hopefully you get the idea.

This concept is called **continuity** – there are no sudden or inexplicable jumps in your path. Think of how disconcerting and weird it would be if, while you were walking past room 367, you suddenly ended up at the top of the Eiffel Tower? Or if the door after 343 was 434 (on a completely different floor), followed by 444 (in the hotel across the street), then finally 344 (back in the hallway and hotel you started in). Continuity basically says that space is smooth and predictable, that to get from one location to another, you have to pass through all of the locations that lie between them.

In addition, you can't do this instantly. To return to the hotel analogy, perhaps walking from room 301 to room 399 takes four minutes at a leisurely stroll. If you jog, perhaps it will take three minutes, and at a full run it takes two minutes. But can you do it faster? Perhaps on a motorcycle it would only take 60 seconds. If we could somehow mount up a rocket sled on tracks, maybe we could get the time down to a few seconds[20]. No matter how good your rocket, however, you cannot get to the end of the hallway instantly. It will always take some amount of time, however small, to traverse the distance.

On top of all that, imagine stepping out of your door and running for two minutes in the wrong direction! In the best-case scenario, you'll end

20 Let's ignore for now the whole "wanting to remain alive" issues that can crop up when dealing with rockets.

up with a bruised nose from the wall; if you're less lucky you may end up plummeting three stories to the sidewalk! Wait, back up – let's at least imagine that the pool is there! The point is that the direction matters as well. Motion depends not only on changing your location, but also on the direction of the change[21] and the time the change takes.

So really, a pretty good definition for motion is:

> **motion** (*noun*) – Change in location over some amount of time in a given direction. Another word for motion is velocity.

But even this definition of motion a bit tricky. Have you ever sat in a moving car and watched the scenery go by? Is the car moving or the scenery?

The correct answer is: it depends. After all, consider the Earth moving around the Sun. Or, is it that Sun moves around the Earth? For centuries, scientists and philosophers believed the latter, based upon the available evidence. What about the motion of the Sun around our galaxy or the motion of our galaxy among the Local Group? How can we reconcile all of this potential movement?

It all depends on what we choose to be our frame of reference.

Frame of Reference

You can think of the frame of reference as being like a picture frame. To figure out where things are in a painting, you measure from the edges – from the frame. Effectively, the frame determines how you measure loca-

21 Lewis Carroll (aka the mathematician Charles Dodgson) summed up the issue of direction quite well using this exchange from Alice in Wonderland:

"Cheshire Puss," she began, rather timidly, as she did not at all know whether it would like the name: however, it only grinned a little wider. "Come, it's pleased so far," thought Alice, and she went on, "Would you tell me, please which way I ought to walk from here?"

"That depends a good deal on where you want to get to," said the Cat.

"I don't much care where -" said Alice.

"Then it doesn't matter which way you walk," said the Cat.

"- so long as I get somewhere," Alice added as an explanation.

"Oh, you're sure to do that," said the Cat, "if you only walk long enough."

tion – and, by extension, motion. If something is at rest, it's at rest relative to the frame – its location isn't changing based on where you're measuring from. This means that the frame is a choice!

So in this case, we can resolve all of the previous examples as being examples of different frames. We often choose the Earth as our frame of reference just because it's so big. However, when we consider orbital mechanics, the Sun is usually chosen as our reference frame. When you're riding in a car, you could choose the car's interior as your frame of reference. If you did, then you're at rest, and it's everything else – the road, the trees, the Earth, the Sun, the Moon, and the stars – that is moving!

Now, it may be quite tempting to simply always choose yourself as your frame of reference, simply so you can legitimately say that everything moves around you. However, you may want to reconsider. Some frames of reference make it easier to solve problems, some make it difficult, and some can make it downright impossible! So while it's nice to know that you get to choose your frame of reference, remember to think of it as a tool to make your life easier.

Forces: Newton's Second Law

Remember Newton's first law? A quick review is in order, if simply as a good introduction to Newton's second law. Newton's first law states:

An object at rest will remain at rest and

an object in motion will continue that same motion

unless the object is affected by a net external force.

Remember that rest and motion are based upon whatever frame of reference you choose. So another way of writing Newton's first law is:

An object's change in motion results

from a net external force.

To really understand this statement, however, we need to dive deep again. Let's start with changes in motion – a.k.a. **acceleration**.

Acceleration

Return, in your mind's eye, to your imaginary car. (It's all repaired now, if necessary – that incident with the deer was back at the beginning of the previous chapter.) Imagine sitting at an intersection, waiting for a red light to turn green. The car is at rest relative to the ground. (The Earth will serve nicely as our frame of reference for this thought experiment.) When the light turns green, your foot presses the gas pedal, initiating a chain of events that ultimately starts your car moving forward. You press the pedal harder, and the car speeds up – it's not just changing locations, it's changing locations at a faster and faster rate! This process is called acceleration. Obvious, right?

Okay, now imagine that your car approaches the light at the next intersection. Oh, it's turning yellow, and now it's red. Your foot lifts from the gas pedal and presses the brake. Another complex chain of events occurs, and the car begins to slow down. You continue pressing the brake and the car moves slower and slower, until eventually it stops at the next light. You're at rest again, but in a different location. What is this process of slowing down called?

If you answered "deceleration," you've fallen into the trap of popular culture. In physics, **any** change of motion is an acceleration. If you speed up, you're accelerating; if you slow down, you're accelerating; if you turn a corner or otherwise change direction, guess what? You're still accelerating!

This last point is a subtle and important point, and probably one of the most confusing aspects of dealing with physics. Acceleration doesn't just measure how much your motion changes; it also measures the direction your motion is changing. The picture on the facing page illustrates a pretty basic situation: an object starting at rest undergoes an acceleration.

The picture on the opposite page is an excellent illustration of the first example above. You are stopped at a red light. When the light turns green, you press the gas. The car isn't moving yet, but the engine is working to change that. (In other words, the car's motion is changing – it's accelerating!) After a little bit of time, the car is moving forward slowly, but your foot doesn't let up from the pedal. You're still speeding up. A little bit later, you're moving faster. But, it's not fast enough, so you keep your foot pressed down. The longer you accelerate, the faster you go.

On that picture, the black arrows represent your motion – how fast you're moving at that given point in time, and the direction. The grey arrows represent your acceleration – how quickly your motion is changing. There are a couple of things to note in this picture. The first is obvious: The black arrow is getting longer, while the grey arrow stays the same. Remember, acceleration measures change in motion. The black arrow gets longer specifically because the grey arrow is there. The bigger the grey ar-

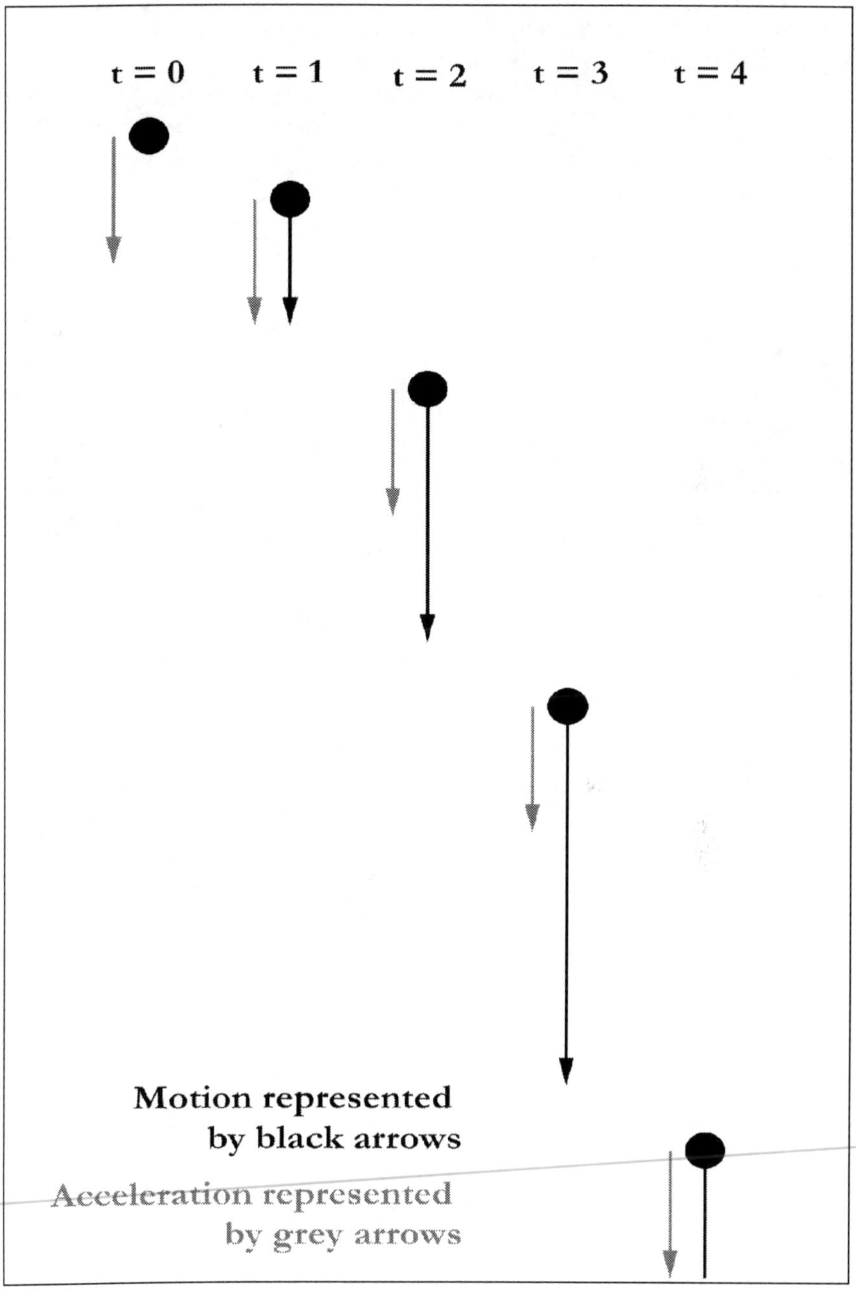
t = 0
t = 1
t = 2
t = 3
t = 4
Motion represented by black arrows
Acceleration represented by grey arrows

row, the more the black arrow should change as the time changes. If you compare the black arrow from one point in time to the next, the amount it has grown is exactly equal in size to the length of the grey arrow. Literally, the acceleration (the grey arrow) is changing motion (the black arrow)! Because they're both pointing in the same direction, it's easy to see how everything adds up.

Imagine, however, that you were moving down a hill and your engine cut out. Some superhero swoops in to save the day, and begins pulling on your bumper. The image on the facing page is a pretty good top-down view of what happens in that situation. At first you're racing down the hill, but the superhero is slowing you down. She's changing your motion, but it's in the opposite direction of how you're moving. In the image, the acceleration (the grey arrow) and motion (the black arrow) are in opposite directions. Over time, the acceleration effectively cancels out part of the motion, effectively subtracting itself out to make the motion smaller. In fact, if the acceleration continues even after the motion is gone (the superhero keeps tugging on your car even though you've come to a stop), then your motion will continue to change. Your car will start moving the opposite direction and begin heading back up the hill due to that superhero's efforts!

If you compare the previous two images, the acceleration was the same in both pictures. It pointed the same direction (down the page) and was the same size. The effect it had on the motion was also the same in each case – it made it bigger downward. The reason it looks so different in the two images is because the initial motion was different. In the first, the initial motion was zero; the dot was at rest, so the acceleration increased the motion downward. In the second image, the initial motion was up the page. The acceleration first had to decrease the motion upwards until it was gone, then it could increase the motion downwards. This is why there's no such thing as deceleration in physics. Anything that reduces an object's speed is just an acceleration pointing in the opposite direction.

Thus, a good definition for acceleration is:

> **acceleration** (*noun*) - Change in motion over some amount of time in a given direction.

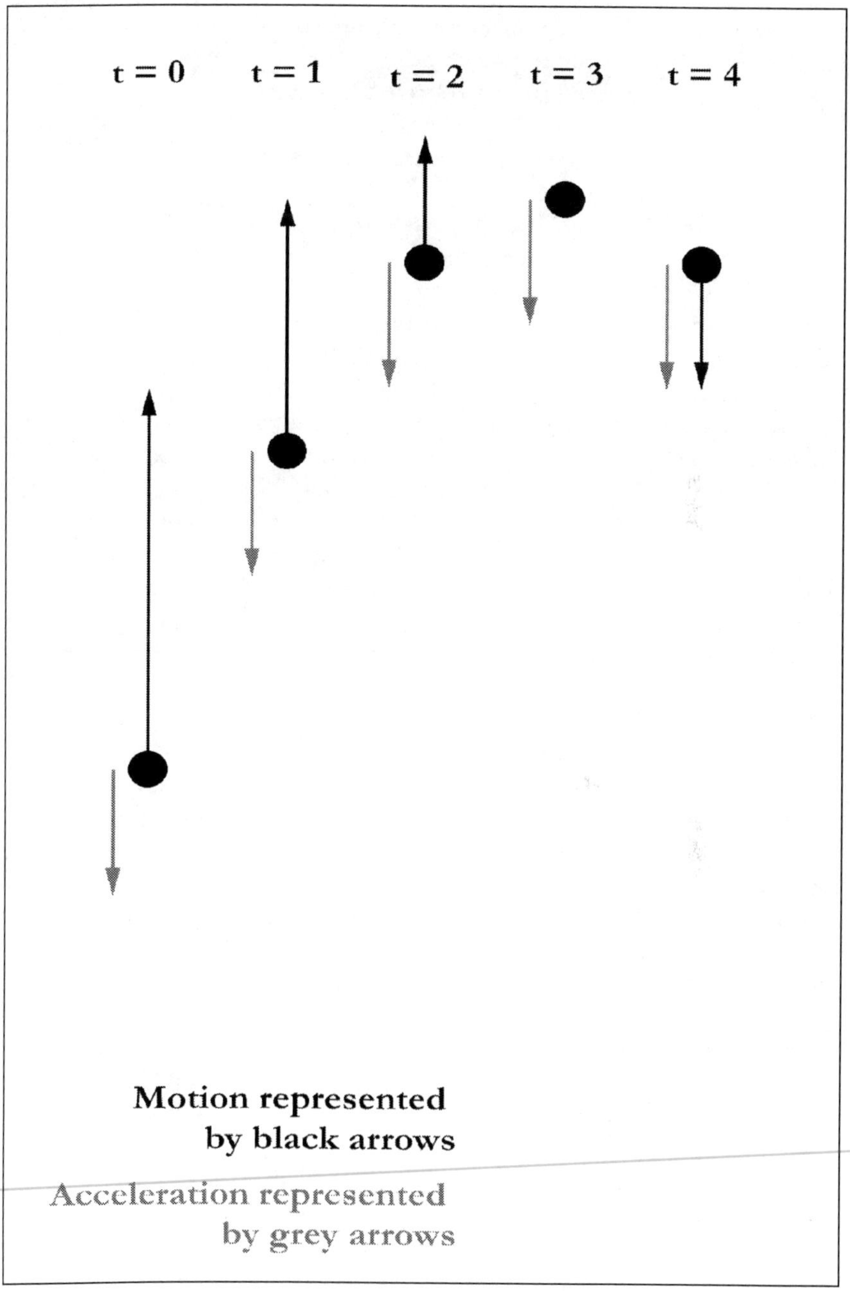
t = 0
t = 1
t = 2
t = 3
t = 4
Motion represented by black arrows
Acceleration represented by grey arrows

So, to restate Newton's first law using our new language,

Acceleration of an object results from a net external force.

But we still have a piece missing! How much acceleration do we get from a given net external force on a given object? Think about pushing on your little brother with all your might, versus pushing on your car with all your might. The change in motion that results will probably be quite different. Why the difference? A little thing called **inertia**.

Inertia measures how much force it takes to produce a given acceleration – the larger the inertia, the larger the force necessary to make the object change its motion by a given amount. If it sounds like I wrote this definition based upon the question I asked above, it's because I did. The concept of inertia is literally the connecting glue between acceleration and force, and it's impossible to define inertia, or even to talk about it, unless it's within the context of acceleration and force. Things get even sticker, however, when I bring in the actual measure of inertia, which is mass.

Gravitational Mass vs. Inertial Mass

We encountered mass before when we talked about weight. That kind of mass – the kind that hooks into the gravitational field to determine how much it pulls on you – should have absolutely nothing at all to do with this kind of mass – the kind that determines how much a force will change your motion. They measure completely different properties of matter, and therefore should be completely independent. Actual measurement, however, shows otherwise. No experiment has ever detected a difference, down to one part in a trillion. For comparison's sake, this is the equivalent of measuring the diameter of the Earth and looking for a difference the size of a single red blood cell.

This issue of gravitational mass versus inertial mass is one of the great mysteries of physics. Some physicists will tell you that this mystery is solved and that Albert Einstein proved they are the same. This is not, in fact, true. Instead, the theory of general relativity *starts* with the assumption that the gravitational mass and inertial mass are always the same, and proceeds from there. To date, I have never heard a satisfactory explanation of *why* they should be identical, other than something along the lines of "they have to be."

So, really, what is mass?

Honestly, that's a great question. Physicists don't really know what mass is, so we usually call it "a property of matter[22]." We can, however, describe what effect it has. It interacts with the gravitational field to produce weight (as we discussed in the previous chapter). For some inexplicable reason, it also relates acceleration to force, which returns us to the topic of Newton's second law.

Back to Newton's Second Law

Based upon all of this, we can actually write a much more thorough form of Newton's first law:

The acceleration of an object is directly proportional to the net external force applied to that object, and inversely proportional to the object's mass.

Or,

$$\vec{a} = \frac{\sum \vec{F}_{ext}}{m}$$

22 Of course that begs the question, "What is matter?" I'm not going there in this chapter, or in this book, and probably not ever. *cough*waves*cough* You could also ask "What's the matter?" to which I would respond, "Sorry, I may be catching a cold."

You can probably guess that a stands for acceleration, m is mass, and F_{ext} is for external force. The Σ is a mathematical symbol meaning sum, and is there to remind you that you have to add up all of the forces on the object to get the net force. This equation is called **Newton's second law**; in reality, it's just a restatement of Newton's first law – the equation can literally be read as "the net external force"

$$\vec{a} = \frac{\boldsymbol{\sum \vec{F}_{ext}}}{m}$$

"on an object"

$$\vec{a} = \frac{\sum \vec{F}_{ext}}{\boldsymbol{m}}$$

"determines the acceleration."

$$\boldsymbol{\vec{a}} = \frac{\sum \vec{F}_{ext}}{m}$$ [23]

When it comes to force problems, almost all of the work involves preparation for applying the equation form of Newton's second law to that problem. Once we've written an equation for the acceleration, we're effectively done with the physics.

Now, you might be wondering why I didn't write Newton's second law using the more familiar form:

$$\sum \vec{F}_{ext} = m \cdot \vec{a}$$

23 If you can learn to read your equations just like they were sentences, you will have conquered one of the biggest stumbling blocks to learning and using physics. In addition, you will have made a huge leap in your mathematics fluency. You can read more about this skill in the appendix "Philosophical Musics and Geometric Reminders" on page 239.

There is one very basic reason for this. Forces determine accelerations, not the other way around. This form of Newton's second law obscures this fact. In almost every force problem, you'll be examining the situation to figure out what the forces are. Then you'll add all those forces together, while accounting for their different directions, to find the net force. Finally, you'll use the net force and the mass to figure out the acceleration. Even in situations where you're given the acceleration, that acceleration was due to a force. Forces are *never* the product of an acceleration (which is what the "familiar" form implies)[24].

Now, let's see how to pull this all together to actually solve problems!

24 I trust you to do the algebra if at some point you need to calculate the net force that produced a given acceleration.

Forces: Conceptual Scaffold

Finally! After all that background and preparation, we're ready to solve problems! Here are the steps of the conceptual scaffold for force problems:

1. Prepare
2. Draw a picture
3. Identify the forces
4. Choose a coordinate system
5. Break forces into components
6. Apply Newton's second law
7. Do math and solve

The rest of this chapter explains each step in more detail; in the next chapter, I'll show you the conceptual scaffold in action by applying it to several different sample problems.

Before we jump into specifics, however, I just want to remind you of a couple of things. First, the conceptual scaffold isn't a solution - just like in construction, scaffolding isn't a building. Although the conceptual scaffold will give you a set of steps to follow, you'll still need to use your knowledge and understanding to make decisions regarding the specific problem you're trying to solve. That's why I spent so much time talking about forces, motion, mass, and acceleration in the previous chapters. That information will (hopefully) help to inform and to guide your decision-making.

Second, every step of the conceptual scaffold is designed to use what came before to prepare for what lies ahead. To successfully complete a given step, you will need to have completed the previous steps correctly; successfully completing a given step means that you're prepared to successfully complete the steps that lie ahead. To return to the construction analogy, you can't build the third floor before you build the first and second, but once you've built the first and second floors, you're ready to tackle the third.

The last step, "Do math and solve," is both the most important and least important of all the steps. It is the one where you finally answer the question that was posed. However, that is also the only step of the scaffolding that does not actually involve any physics. Once Newton's second law has been successfully set up, all that remains is to solve a system of equations to find the aforementioned answer. If you know someone who's taken algebra (or, for you calculus-based folks, calculus or differential equations), give them the equations you found and ask them to solve for the variable of interest. They should be able to do it without knowing the first thing about physics.

One last thing that I should mention: If you're in a traditional physics course, you've probably been wondering where the kinematic equations are. You'll find them in a section named "Kinematics" at the end of the scaffolding in the "Do math and solve" section. It's where they really belong.

1. Prepare

Before the first brick is laid or the first scoop of dirt is dug, before the architect even considers a rough sketch, a lot happens: Resources are prepared, the site is surveyed, and needs are assessed. Similarly, before you start building your solution, you should do a few things to get ready.

First, relax. Take a deep breath, hold it for a second, then let it out[25]. Like a runner at the beginning of a marathon, prepare yourself mentally for the work ahead. Make sure you have all your tools ready. You may want to use pencil instead of pen, and have an eraser handy. (Not that I think you're going to make any mistakes, but, you know, just in case.) If you want scratch paper it should be nearby, although I would suggest that you

25 Depending on the situation, previous experiences, and your overall temperament, you may have to do this a few times.

do all of your work, including scratch work, on the same paper with your regular work. If you're doing practice problems or homework, you'll want everything easily available for review. If it's an exam, then your professor should be interested in all of the steps you took, for good or for ill. Some of my students used color pencils (or, in one case, crayons[26]) for their drawings.

Now, read the question. Once you've read it, take a moment to imagine the situation described and study any picture provided. Then, read the question again. If there is anything you don't understand, ask about it. Getting clarification is not the same thing as asking for the answer. If necessary, take a moment to relax again.

Finally, you're ready to begin. You can do it!

Looking Forward: Make sure you understand the situation as described in the problem statement. Your next task is to draw a picture of that situation, so if you're having trouble imagining what's going on, you should get more information.

2. Draw a picture

Now it's time to put your drawing skills to the test. It doesn't have to be high (or even low) art[27]. The purpose of this picture is to act as a reference and reminder of the physical situation the problem statement describes. It will serve as a foundation for the steps that follow. Now, look at your picture. Which object(s) are the most important, as described in the problem? Review your picture and the problem statement. Which object will change the way it moves, either by slowing down or speeding up? Failing that, which object is being constrained and shouldn't be moving? That object is most likely your object of interest. (In a few cases, you'll have multiple objects of interest tied together – sometimes literally.) If the ob-

26 In an exam situation I suggest you check with your instructor before breaking out a box of Crayola's finest. (Crayola is a registered trademark of Crayola,LLC.)

27 In the chapter of examples that follows, you'll see my drawings. Please don't laugh.

ject you've identified is not front and center in your picture, you will probably need to redraw it.

Looking Back: Reread the problem statement and make sure that all of the key elements of the problem are represented in your drawing. Your object of interest should focus prominently in your drawing.

Looking Forward: Later on you will be drawing forces on the picture you've drawn, so make sure the picture is big enough to easily accommodate those additions.

3. Identify the forces

The next step is to figure out what forces are acting upon the object(s) you identified in the last step. Anything that pushes or pulls on that object should be written down. In addition, you'll need to figure out which direction each of those forces points. Once that is done, draw all of the forces on the picture, making sure that they are obviously acting upon the appropriate object in the correct directions. (The easiest way to do this is using arrows.) At this point, you have built a force diagram - a precursor of the free body diagram that most physics professors love to see. As long as your diagram is correct, you don't actually have to make a free body diagram. Having said that, it never hurts to impress your instructor!

To make a free body diagram from your picture, you'll need to start with a new drawing for each object of interest. Draw a big dot to represent your object of interest. Then, redraw all of the forces you had on that object out of that dot. (In other words, the tail ends of the arrows should meet at the dot.) Make sure the arrows still point in the same directions! Do this for each object of interest, and voila! You have a set of free body diagrams. Personally, I like the force diagrams better, but you should use whichever floats your boat and makes the problem easier to solve.

Looking Back: Look at the picture and read the problem statement (again). Anything that is directly touching the object might be exerting a force upon it; in addition, there may be other forces acting either at a distance (field forces, like weight or electric force) or from gaseous objects (like thrust or drag). You may need to make reasonable assumptions. Remember, if the situation described occurs near the Earth, it's reasonable to add the object's weight. However, do *not* include any forces that stopped acting *before* the situation occurred or any forces that do not act until *after* the problem is finished. For example, often you will not need to worry about the force that launches a ball into the air to give it an initial velocity, or the force of the ground acting upon the ball after it lands. Check out the sample problems to see what I mean.

Looking Forward: Looking over your forces, consider their directions. Is there any direction that seems special? Are any of the forces aligned perpendicularly to each other? Keep these directions in mind, as they may come in handy in the next step. Also note that while it may be tempting to write down equations for the magnitudes of each of the forces at this point, it isn't necessary to do it yet. It's okay if you don't know what the magnitude of the force is. Right now, you just need to know its direction.

4. Choose a coordinate system

Because you will eventually be moving from vectors (arrows) to equations, it will be important to have a set of reference directions. These will determine what's positive and what's negative for building your equations. Toward that end, you need to choose a local coordinate system. This is like choosing a frame of reference, except instead of specifying what's considered at rest, you're specifying which direction(s) will be positive. The local coordinate system is most easily set up as a Cartesian set. That

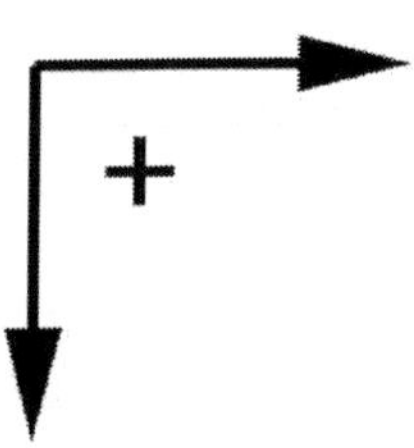

means you'll have one, two, or three special directions (each of which is called a coordinate axis[28]), which are all at right angles to each other. The set you know (and love) is probably x, y, and z; that's also the one your physics professor is probably going to shove down your throat. Just so you know, however, ANY coordinate system will work as long as the coordinate axes are at right angles to each other. It really is your choice. An

28 "Talking of axes," said the Duchess, "chop off her head!"
– Alice in Wonderland, Lewis Carroll

example of a perfectly valid coordinate system can be seen on the previous page. Although the plus sign isn't absolutely necessary, I find it a useful reminder that the arrowheads are defining the directions I am choosing as positive.

Just like with frames of reference, your choice of coordinate system can make a problem easy, hard, or impossible to solve. As a rule of thumb, it's best to use a system that points one of the coordinate axes in the same direction as you expect the final acceleration to be. This is actually more an art than a science. Practice and patience will improve your ability to guess correctly[29]. However, here a few suggestions that may or may not always work: If it's sliding, then it will probably stay on the surface. If it's not on a surface, then straight up (or straight down) is probably a good choice. If you have no idea whatsoever, then you should try to line up as many forces as possible with coordinate axes. Remember, to count as a coordinate system the coordinate axes have to stay at right angles to each other, so this method only works well when most of your forces are also at right angles to each other.

The labels for each of the coordinate axes are purely for reference, and you may use whatever you like. As I said, physicists always tend to use x and y (and z) by training. I personally prefer to go with something more interesting: names ("Buzz" and "Woody"), related sets of words ("Hot," "Jelly," and "Doughnut"), or stupid jokes ("Exe" and "Why"). There's no rule that says physics has to be boring!

<u>Looking Back</u>: Review all of the forces you've labeled and make sure you haven't forgotten any. You are choosing your coordinate system so that you can translate your force diagram into equation form, so make sure you've taken all of the forces into account.

<u>Looking Forward</u>: Again, the purpose of the coordinate system is to make sure that your equations match your diagram and make sense. Choosing an appropriate coordinate system now should make the next step easier.

---

29 I like to call the ability to guess correctly *intuition*. It's three parts experience mixed with one part logic and two parts "Oh, why not?"

5. Break forces into components

WARNING! We now interrupt this physics discussion for a special math announcement!

Remember when I said that the last step was math and everything before that was physics? Yeah, well, I may have exaggerated that slightly, as this step will bear out. If you're lucky, all of your forces point in one of the special directions of the coordinate system you chose in the previous step. In this case, this step will be ludicrously easy. If, however, you've gotten a typical physics problem, then you'll need to break your forces into components.

What are components?

Basically, any vector (like a force) can be written as the sum of two other vectors, as long as certain rules are followed: The three vectors must form a triangle, the head of the single vector must meet with the head of first vector of the pair, the tail of the single vector must meet with the tail of the second vector of the pair, and the tail of the first vector of the pair must meet with the head of the second vector of the tail. Whew! Writing that sentence made my head hurt, and I'm familiar with vector addition! I don't even want to think about what it was like reading it. Instead, check out the picture here.

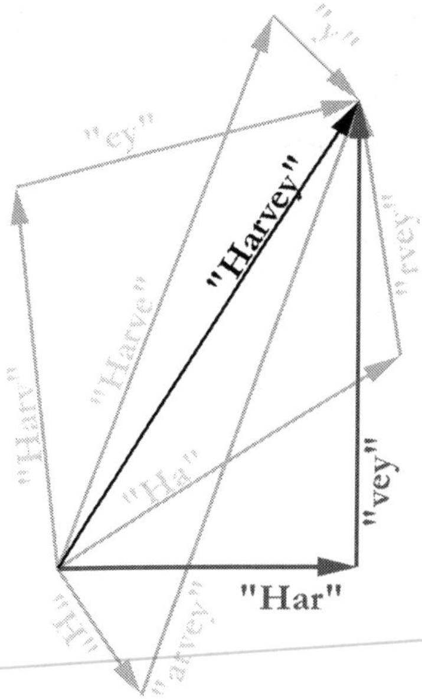

In black is the vector that I've named **Harvey**. Say hello to **Harvey**. **Harvey** says hi. Now that you've been properly introduced, we're going to break **Harvey** up. The grey arrows are some examples of how we can break up Harvey. Notice how the grey arrows naturally form up into pairs – **H** and **arvey**, **Ha** and **rvey**, **Har** and **vey**, **Harv** and **ey**, and **Harve** and **y**[30]. Each pair meets head-to-tail within themselves (so the head of **H** meets the tail of **arvey**, for example), the tail of each pair meets **Harvey**'s tail, and the head of each pair meets **Harvey**'s head. (If these were driving directions, then **Harvey** would be the direct route, and each of the pairs would be alternate routes.) As long as the three form a triangle, and generally show the same directionality ("head-to-tail" direction), it works. Out of all of those, however, I've chosen one pair to be a darker color of grey – **Har** and **vey**.

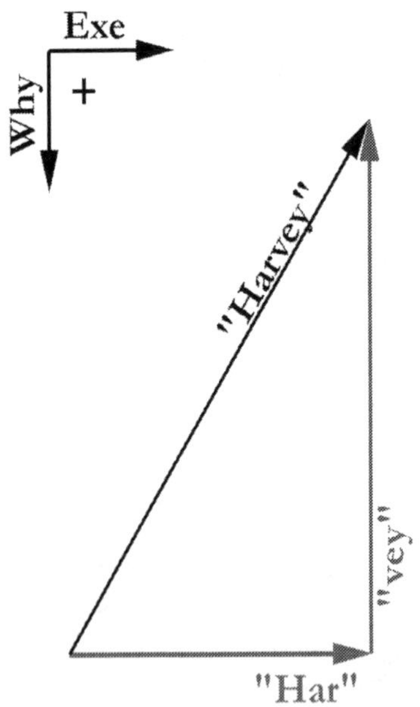

This whole vector addition thing means that you can break any force vector up into pieces that are parallel to the special directions of the coordinate system that you chose in the previous step. For example, for **Harvey**, **Har** and **vey** would be components if the "Exe" direction is to the right, and the "Why" direction is downward. **Har** would point in the same direction as "Exe" (or the "positive Exe" direction) and **vey** would point in the opposite direction of "Why" (or the "negative Why" direction). Note that it doesn't matter if the vectors point in the same direction or opposite direction as the coordinate axis, so long as they are parallel. As you can see in the example above, whether they point in the same direction or opposite directions only influences the sign. I'll tackle that a little bit later, however – in the next step.

Your work has only started once you've drawn out the components. You need to relate those pieces back to the original vector! In other words, you need to figure out the relationship between the size of **Har**, the size of **vey**, and the size of **Harvey**. Guess what this means? More math – spe-

30 **Harvey** says, "I see what you did there!"

cifically, geometry and trigonometry[31]. As it turns out, there's only one bit of information you need to know from trig: **SOHCAHTOA**.

This is the point when I would ask the students in my physics classes to repeat it out loud with me with the same voice and tone they might use to recite a mantra or calming phrase. I'll spare you this exercise[32]. Just make sure you remember it. SOHCAHTOA!

If you haven't had trig, you're probably wondering what the heck SO-HCAHTOA is. The short explanation is that it's an acronym to remember the relationship between the sides of a right triangle and the angles between those sides:

Sine
Opposite
Hypotenuse
Cosine
Adjacent
Hypotenuse
Tangent
Opposite
Adjacent

In other words, if you choose one of the smaller angles in a right triangle (let's call that angle *Bob*), the sine of *Bob* is equal to the length of the side directly across from *Bob* divided by the length of the hypotenuse (the longest side of the triangle). The cosine of *Bob* is equal to the length of the side adjacent (or next to) *Bob* divided by the length of the hypotenuse, and the tangent of *Bob* is equal to the length of the opposite side divided by the length of the adjacent side.

31 And they assured you that this was "algebra-based physics"! (If you're taking calculus-based physics instead, you have no room to complain.)

32 It sounds silly, but it's hard to forget the word once you've stood around chanting it at the top of your lungs. Try it!

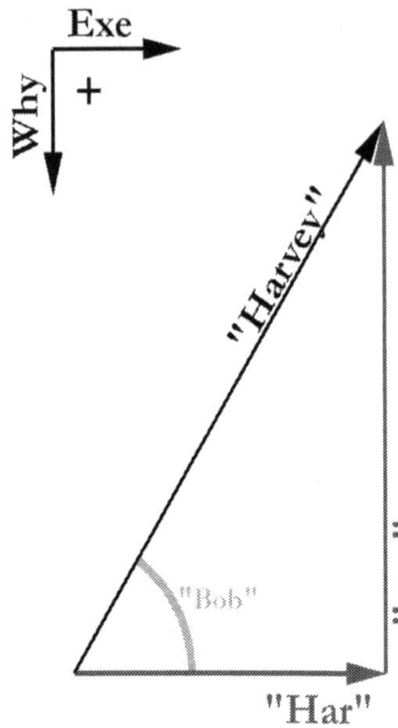

For example, let's introduce **Harvey** to *Bob*. In the adjacent picture, I've marked *Bob* as the bottom arrow of the triangle. (Our choice of angle for *Bob* is arbitrary; we could have just as easily made *Bob* the upper angle – the one between the arrowheads – instead.) Note that the angle between **Har** and **vey** is 90° – it has to be, because we based **Har** and **vey** on the coordinate system we chose in the last step – "Exe" and "Why." Because this is a right triangle, we can use SOHCAHTOA.

The longest side of the triangle is **Harvey** – that's the hypotenuse. The angle *Bob* is formed by **Harvey** (the hypotenuse) and **Har**, so **Har** is the adjacent side (it's next to *Bob* and not the hypotenuse). The only side left, **vey**, is on the other side of the triangle from *Bob*. It's the opposite side. So, based on SOHCAHTOA, the sine of *Bob* is equal to the length of the opposite side (**vey**) divided by the length of the hypotenuse (**Harvey**):

$$\sin(Bob) = \frac{vey}{Harvey}$$

The cosine of *Bob* is equal to the length of the adjacent side (**Har**) divided by the length of the hypotenuse (**Harvey**):

$$\cos(Bob) = \frac{Har}{Harvey}$$

The tangent of *Bob* is equal to the length of the opposite side (**vey**) divided by the length of the adjacent side (**Har**):

$$\tan(Bob) = \frac{vey}{Har}$$

At this point, I have some good news and some bad news for you. The good news is that you're almost done with this step (YAY!). The bad news is that you still don't know how big *Bob* is. Ten degrees? 20? $\pi/6$ radians? You may have been given an angle in the problem statement, but that might not be *Bob*'s size[33]. Unfortunately, there's no sure way to figure out if it's *Bob*'s size (or, if it isn't, what *Bob*'s size actually is) without using some principles of geometry. I've given examples of the ones that most commonly crop up (for physics problems, at least) on page 243, in the appendix "Philosophical Musings and Geometry Reminders." I also suggest you look over the sample problems to see this process in action.

<u>Looking Back</u>: Check to make sure you have accounted for all of the forces you identified in the previous steps. Each one should either lie entirely along one of your chosen coordinate axes or should be broken up into components.

<u>Looking Forward</u>: Note for each force's component (including the forces that lie entirely along one coordinate axis) whether it points in the positive direction of the coordinate axis or the negative. These signs will be used when using Newton's second law in the next step.

6. Apply Newton's second law

Now it is finally time to build our equations based upon our understanding as represented by the work we've done in the previous steps. At this point, if you've done everything correctly, you should be able to finish up the actual physics.

---

33 Please don't try to force *Bob* into something that doesn't fit him. There will be a huge ripping noise, and not enough brain bleach exists to erase the resulting mental picture. Make sure the angle fits!

If you'll remember from the previous chapter, Newton's second law is

$$\vec{a} = \frac{\sum \vec{F}_{ext}}{m}$$

As I said when I first introduced Newton's second law, once you've written an equation for the acceleration, you're basically done with the physics. Everything you've done up to this point has been in preparation for this task: finding the acceleration.

So, now what do you do? Well, there are three pieces of Newton's second law: the acceleration, the mass, and the sum of the forces. The acceleration and the mass are pretty simple; either you know them or you don't. The real work, then, comes in figuring out how the forces add up.

Because acceleration is a vector quantity (like force), it can also be broken up into components (like force). Conversely (and this is the useful part for you), if you find the components of the acceleration, you can combine them to find the acceleration as a vector quantity. Therefore, when you apply Newton's second law, you'll be applying it for each direction of your chosen coordinate system, for each object of interest. In other words, you'll go through each object experiencing forces, and for each direction, you'll add up the force components for that direction, taking any component that points in the same direction as the coordinate axis as positive, and any that points in the opposite direction as negative. You'll then divide that total by that object's mass to get the acceleration in that direction. Once you have the components of the accelerations for every object, you can find the net acceleration as a vector quantity for each object.

Or not.

Much more likely, what you'll actually want to find is a distance, as the question asks "how far <something something>." Or you might need to find a speed ("how fast") or a time ("how long"). All of these require that you find the acceleration(s), but only as an intermediate step. Just as acceleration and force are vector quantities, distance and speed can be written as vector quantities[34]. As with other vector quantities (such as acceleration and force), it's easier to work with these in components as well. In addition, depending on the problem and your choice of coordinate system, you may be able to specifically state that one (or more) of the components of the acceleration is zero[35]. In all cases, it is almost[36] always better to work with the components of the acceleration.

Don't worry yet about actually solving the equations you get; that thankless task is tackled in the next step. For now, just write down your equations and bask in the fact that you've actually done some physics!

Looking Back: If any of the forces (or components) you have from the previous steps does not line up along one of the axes of your coordinate system, then you must go back to Step 5 and break that force up into components. You may also be able to set some or all of the acceleration components equal to zero, depending on the problem and on the choices you made earlier.

Looking Forward: At this point, it would be most useful to write down any other equations you think you might need, such as those for finding the magnitude of the forces you identified back in Step 3. If you find that you are missing one or more forces, you will probably need to start over completely[37]. If you find that you have one or more extra forces, you may be able to simply set the magnitudes for those forces equal to zero; however, you should still examine your diagram and make sure that the rest of the forces you identified in Step 3 are correct.

7. Do math and solve

If everything has gone well, then at this point the physics is basically done. All that's left is to tweak the knobs and fiddle with the antenna in

---

34 These are called displacement and velocity, respectively.

35 See the example problems in the next chapter for, well, examples.

36 "Almost" in this sentence carries the same weight as it does in the sentence, "It is almost always better to have oxygen to breathe than not." In both cases, the exceptions are *very rare*.

37 Go directly to Step 1. Do not pass "Go"; do not collect $200.

the hopes that eventually a coherent picture will appear on the fuzzy screen. Wait, that's the procedure for getting an old TV set to work. Kick the tires? Manipulate the equations? Oh, that's the one.

Seriously, though, it is now time to literally "Do math and solve" for your variable of interest, using the equations you built in the previous steps. For algebra-based physics, this will (predictably) involve quite a bit of algebra. I wish I could offer a foolproof and easy way to work with the equations, but basically this step is just a bunch of "plug-and-chug"[38]. There are, however, a few additional equations that you might find useful.

Kinematics

It is extremely rare that a physics problem will ask you to find the acceleration directly. Instead, it will ask "how far did <some object> go?" or "how fast is <that thing> moving?" You may even get "how long does it take for <some situation> to occur?" These can be phrased in a wide variety of ways, but they all boil down to the same three things: distance ("how far"), speed ("how fast"), or time ("how long"). If you've done the first six steps of the forces conceptual scaffold correctly, then you have found an equation based on acceleration. So to answer the question, you need to know the relationship between acceleration (which you have) and distance, speed, and time (which you want). These relationships are represented by the kinematics equations.

Before I simply drop them on you like a ton of gold bricks, however, it's important that I warn you about two things. First, I'm presenting these here "as-is," without any proof or derivation. To properly derive these equations, I'd need to break out the calculus. Oh sure, I could do some fancy mathematical tricks with averages, and "derive" the equations using some algebraic pseudo-calculus, but it would distract you from the real point. The equations are a tool, nothing more. The meaning we assign to the variables is important physics, but once we've made that assignment, everything that's left is simply turning the calculus crank. For this reason, I'm going to simply give you what you need: the equations and what they mean. If you really want to see how they come about, go read the ap-

38 Also known as "turning the crank" and "crunching the numbers".

pendix "Kinematic Equations", starting on page 225. There you'll find them birthed in all their bloody calculus-based glory.

> **For Calculus-Based Physics:**
>
> You guys don't get off this easy, unfortunately. If you really want to succeed in a calculus-based physics class, you're going to need to know all of those gritty details. You should review the derivations that start on page 225, then look over the calculus-based example problem (page 229) to see how this last step changes. All of the other steps, however, will work for you just as well as they work for the algebra-based students. You'll just need to be a little bit more careful, contentious, and kind. (It *never* hurts to be kind.)

The second thing you need to know is that these equations will only work in situations where the acceleration is constant. For a variety of reasons, this isn't as much of a limitation as you might think. The only reason that matters, however, is that to deal with situations where the acceleration varies, one is required to use calculus. In an algebra-based physics class, that's a big no-no. Therefore, all of the problems you'll have to solve will have constant accelerations, by design!

Let me add a huge BUUUUUTTTT[39] here: You may have to tackle problems where the acceleration changes. "Wait," I can hear you ~~whine and moan~~ point out intelligently. "You just said, like RIGHT UP THERE IN THE PREVIOUS PARAGRAPH, that changing accelerations require calculus, and that's a no-no!" And you're right; I did. However, this is one of the sneaky tricks physics teachers like to pull: If the acceleration is a constant value, then changes to a different constant value, you can tackle it as two separate problems – one before the change, one after. Is it fair to expect you to do that? I don't know. In fact, I think I could make a strong case on either side. However, the real question you should be considering is "does it matter if it's fair if I'm expected to do it anyway?"

Thankfully, these kinds of problems are almost always easy to recognize. For example, if the problem statement describes a change in the situation or forces, then it's going to require multiple applications of the conceptual scaffold. Another hint is if the problem has a series of parts that focus on

39 I like big... wait, I made that joke already, didn't I?

a chain of events. An example of this would be "Then, <some event happens>. How long/far/fast <blah blah blah>?" This structure screams multiple accelerations. Just remember that you can literally rebuild the scaffolding from scratch for the second part, just like you did for the first part. You can see an example of this kind of problem in the next chapter.

Okay, now for the kinematic equations:

$$v_f = v_0 + a\Delta t$$

$$\Delta x = v_0 \Delta t + \frac{1}{2} a (\Delta t)^2$$

$$v_f^2 = v_0^2 + 2\text{a}\Delta x$$

In these equations Δx represents the distance traveled (how far), v represents the speed (how fast), a represents the acceleration, and Δt represents the time interval (how long). Any subscript of 0 or i on a variable indicates that it is the initial value. That is, this variable represents the value at the start of the time interval. A subscript of f (or none at all) indicates that it is the final value, or the value at the end of the time interval.

One major change you may notice from the way others write these equations is that I use Δt instead of simply t. This is to remind you that the time represents an interval – how much time has passed from some starting point in time to some ending point in time. Normally this will be trivial, and we'll call the starting point in time zero. If our problem requires multiple applications of the conceptual scaffold, however, then it will be important to track exactly what is the starting and ending points in time for each use of the conceptual scaffold.

In any case, you should work with the symbols in your equations instead of the actual values. Solve for the variable of interest; then, once you're done with all of the algebra, plug in the numbers and do the final calculation. The main reason I suggest this is that it's much easier to figure out where the term μN came from than the value 1.523 if you run into trouble[40]. Otherwise, good luck! You can do it!

Looking Back: If at any time you are unsure of where a particular term in your equation came from, you should be able to track backwards through the steps you used to identify its source. This is the strength of being careful and systematic. Everything should be linked both forward and backward in a logical way. If something comes out differently than you expect, take the time to try to work backwards to figure out why. Once you have an answer to the question, consider whether or not it was what you expected. Does your answer make sense given the problem statement, the picture you've drawn, and the forces you identified? If it does, congratulations! You're done!

40 Trust me; at some point, you WILL run into trouble.

Forces: Examples

Here you will find a few sample problems (and thankfully, their solutions!) using the forces conceptual scaffold. I've set the problem statements here in advance, so that you can try applying the conceptual scaffold by yourself, then check your work against the example solutions.

Example 1: Indrik Linask is helping prepare the stage for the upcoming Estonian Independence Day Celebration by placing speakers around an auditorium. To get the best sound coverage, he needs to hang a 120.0 kg speaker from two anchor points in the ceiling using cords. One of these cords will have to make an angle of 35° with respect to horizontal; the other one will have to make an angle of 55° with respect to horizontal. How much tension will each cord experience from the speaker?

Example 2: Joseph Gascho throws a ball at an angle of 30° up from horizontal from the Empire State Building 86th floor open-air promenade at 1.53 meters per second. The 86th floor promenade is 320. meters high[41]. Ignoring air resistance (and the surrounding buildings), how far away from the base of the Empire State Building will the ball hit the ground?

Example 3: A block is launched up a 30° ramp with an initial speed of 2.25 m/s. The coefficient of friction between the ramp and the block is 0.100. The block slides up the ramp until it stops, then it slides back down to the bottom. How long does it take for the block to make the round trip, and how fast is it going when it reaches the bottom of the ramp?

41 Data pulled from http://www.esbnyc.com/explore_esb_about_esb.asp: "1,050 feet to the 86th floor Observatory" which corresponds to 320. meters.

Example 1

Indrik Linask is helping prepare the stage for the upcoming Estonian Independence Day Celebration by placing speakers around an auditorium. To get the best sound coverage, he needs to hang a 120.0 kg speaker from two anchor points in the ceiling using cords. One of these cords will have to make an angle of 35° with respect to horizontal; the other one will have to make an angle of 55° with respect to horizontal. How much tension will each cord experience from the speaker?

1. Prepare

Normally you should work with pencil and paper. However, I'm recording this work for posterity, so I've got my word processor and image manipulation programs ready. As I pointed out before, it's good to take a deep breath and relax[42]. We can do this!

2. Draw a picture

There are only three things of significance mentioned in this problem: two cords and a speaker. Therefore, our picture will include those three things. There's also Indrik and the auditorium, but these seem like window dressing – basically details that have been added to motivate the problem. How do I know that Indrik and the auditorium aren't important? Well, one clue is that we weren't given any details about them, whereas the problem was quite specific about the speaker and the cords. In addition, the problem statement specifically asks for the force (tension) exerted upon each of the cords by the speaker. These two clues together leave me feeling confident that we only need to deal with the cords and the speaker.

Which one of these, however, is the object of interest? Experience tells me that the speaker is the one we want. By experience, I mean that problems where a cord or rope is the object of interest are extremely rare, even in calculus-based physics courses. In algebra-based courses, they're effectively non-existent. But how might you figure out the object of in-

42 Don't panic! You know where your towel is, right? RIGHT?!

terest if you don't have tons of experience? For that, we'll have to look forward to step 3. For now, however, I'll just draw the picture with the elements I believe to be important.

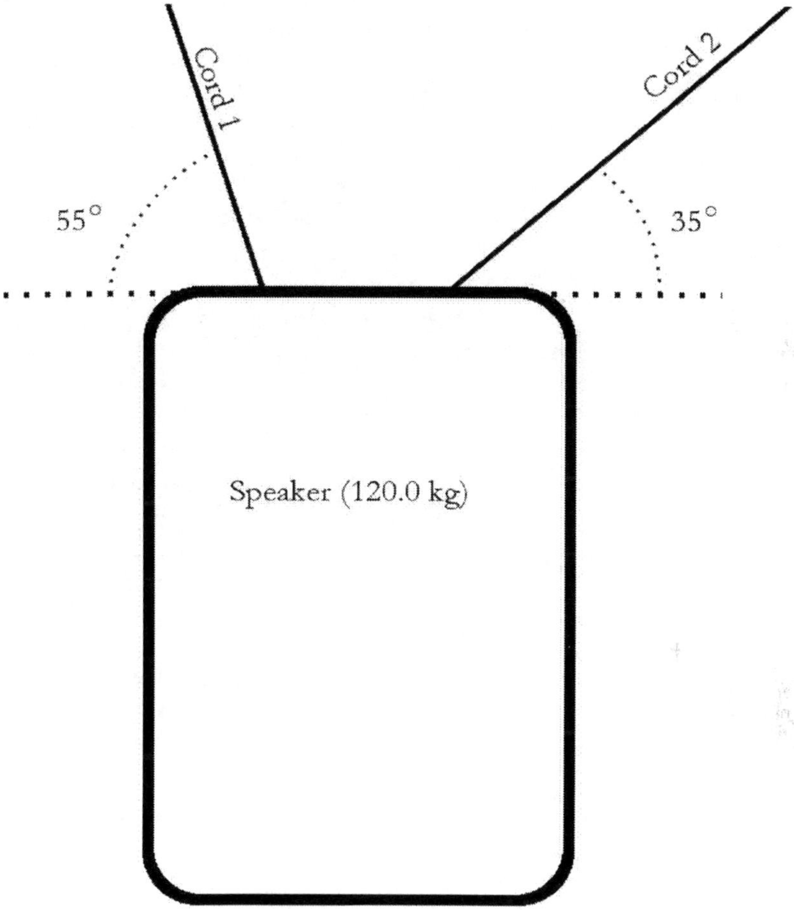

A few things that I'd like to point out about the picture. First of all, it's not very fancy. Simple line drawings are fine, as long as they accurately represent the situation described. Take a moment to look over the problem statement. Does this picture match what was described? Second, I'd like you to notice that I labeled everything in the drawing. That way, if we take a break (or, more likely, get interrupted), we'll find it much easier to pick up where we left off. The labels themselves don't really matter, as long as it's clear what they represent. I also added the information from the problem to the drawing. While this isn't strictly necessary, it makes the information readily available. When we're ready to use those numbers, we won't have to go searching around for them. In addition, adding that stuff to the picture can sometimes help if we're having trouble visualizing what the problem is stating. For example, many people have trouble with angles, especially when they're written as "with respect to[43] horizontal" or "with respect to vertical." If you look at the picture I drew, I put in a horizontal line, then marked the angles from that horizontal line to the cords. Now the angles are properly recorded on the picture itself, which will make our lives easier later on. The final thing I'd like to point out is that the picture I've drawn is nice and big. This gives us plenty of room to work when we start adding in force vectors in step 3.

3. Identify the forces

Now I'll add the forces on the object of interest. As I mentioned above, this step will help confirm that we've chosen the correct object to focus upon. Remember, to solve the problem, we need to add up all of the forces on the object of interest. So, what's pulling or pushing on the object of interest?

If you said gravity, congratulations! We can assume with some confidence that this band is performing on or near the surface of the Earth, so we have to consider the gravitational force exerted on the speaker by the Earth. In other words, we need to take into account the speaker's weight. What direction does the Earth pull the speaker? If you answered straight down or some variation thereof, congratulations again! Gravity always pulls objects directly toward each other. On the surface of planets (like the Earth), this direction is almost always straight downwards, toward the center of the planet. In fact, choosing downward as the direction weight acts is rarely a bad idea.

---

43 Fun fact: "With respect to" comes up in physics problems so often that some professors will abbreviate it as "wrt."

Next we should consider what other forces are acting upon the speaker. What about the two cords? Each of them should be exerting a tension force on the speaker. If I add each of these forces to the picture, we get the following:

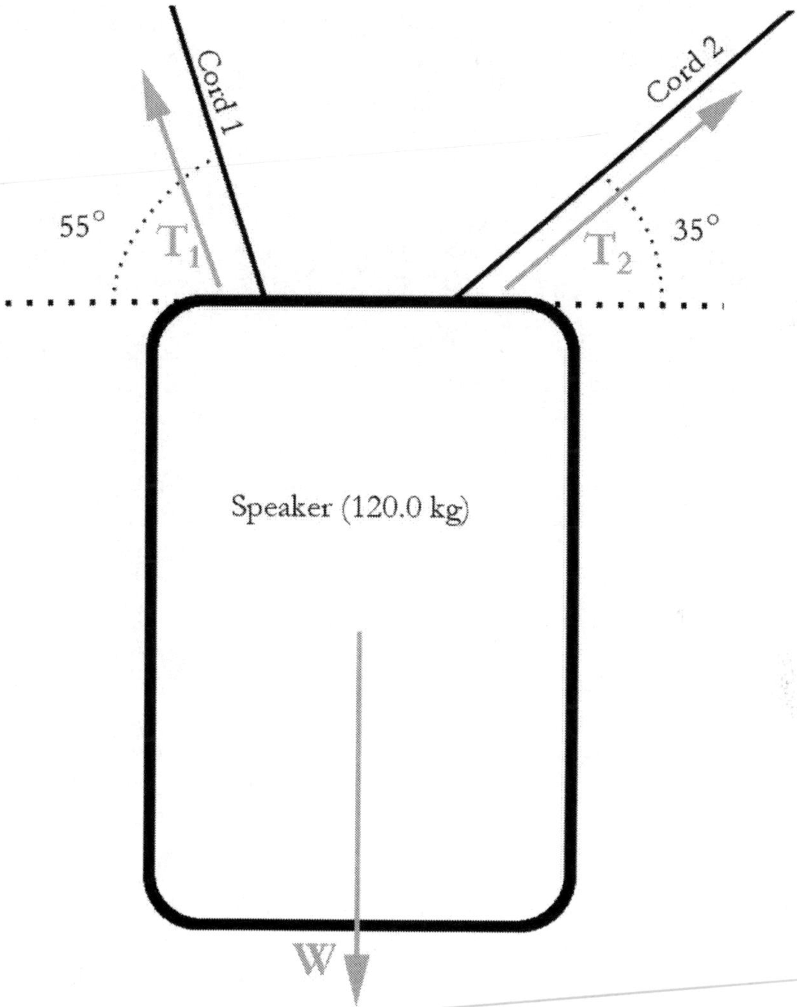

The weight acts on the center of the speaker and is labeled W. Each of the tensions acts on the top of the speaker, and they are labeled T_1 (for cord 1) and T_2 (for cord 2)[44]. We can choose any set of names for our forces that we like as long as we know what each one refers to.

What about other forces that could be acting on the speaker? When we reread the problem statement, nothing else is mentioned, but just to make sure, let's go through the list of common forces. We've already added weight and tension. Normal force requires a surface that presses against the object, as does friction. Elastic force requires a spring or elastic cord; this problem didn't say anything about the cord being stretchy or otherwise elastic. No applied forces were mentioned in the problem statement, nor were any rockets or other source of thrust. We can now be fairly confident that we've found all of the forces acting upon the speaker.

One last thing: I promised we'd make sure that the speaker was the object of interest. So let's pretend, just for this paragraph, that one of the cords is the object of interest. What forces are acting upon that cord? The cord's weight matters. After all, we're still near the surface of the Earth, and gravity still exists. The speaker is pulling on one end of the cord, so there's the applied force from the speaker. Finally, the ceiling is pulling on the other end of the cord, so there's an applied force from the ceiling. However, think about those forces. We don't know the mass of the cord, so we can't figure out how much the cord weighs. The force exerted on the cord by the ceiling will depend on how hard the speaker is pulling on the cord, which means we have to figure out what's going on with the speaker before we can tackle the problem with the cord! So even if we'd chosen one of the cords as the object of interest, it should quickly become obvious that the speaker is where all the action is (so to speak).

4. Choose a coordinate system

So now that we have identified all of the forces, it's time to choose which directions are positive. If we look back at our drawing, there may be a choice that immediately jumps out at you, as shown on the facing page:

44 For now, the direction of each of the forces is much more important than where exactly they are applied. When you do torques and rotations the actual point of application of the force will become more important.

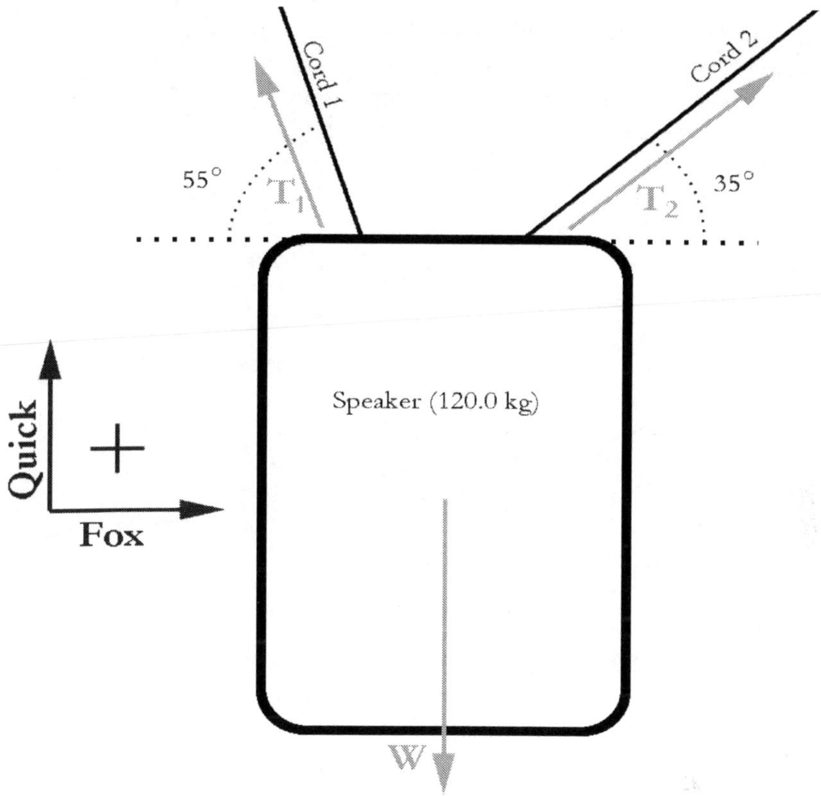

I like this coordinate system. It's an obvious and easy. Let's go with it. The directions are all that matter here. You can name them anything you want. I'm going to go with "Quick" and "Fox."

Now, you might be saying to yourself, "Well, that was easy!" You're right. Choosing a coordinate system is easy. The real trick is figuring out if your choice will make things easier or harder down the road[45]. However, because we're just starting, we're going to work through with these coordinates and see how things turn out. In later examples, however, we'll spend more time thinking about our choice of coordinate system.

45 As it turns out, this choice will make the problem slightly harder than it needs to be. See if you can figure out the less obvious system of coordinates that would make the problem easier. *Hint: think about the angle between the two cords...*

5. Break forces into components

Now that we've chosen our coordinate system, we need to break each of our forces into coordinate components. Let's start with the weight. We said that the weight points straight down; that's exactly the opposite direction as our coordinate "Quick." All of the weight points in the negative (opposite) "Quick" direction. None of it points in the "Fox" direction (either negative or positive). We can make a note of that as follows:

"Quick": **–W**
"Fox": 0

This little list will remind us of what we've figured out so far. What about the two tensions? For those we will need to use SOHCAHTOA.

Because we have the angles between the horizontal and each cord, we can break each tension up directly. Let's start with cord 1. Looking at the picture on the facing page, we can see the triangle formed based upon the coordinate system we chose. The hypotenuse (long side) of that triangle is T_1. The vertical leg of that triangle points in the same direction as "Quick," so let's call it $T_{1+Quick}$. The horizontal leg is in the opposite direction as "Fox," so let's call that one T_{1-Fox}. Using SOHCAHTOA gives us the following three relationships between the sides and the angles:

$$\sin(55^o)=\frac{T_{1+Quick}}{T_1}$$

$$\cos(55^o)=\frac{T_{1-Fox}}{T_1}$$

$$\tan(55^o)=\frac{T_{1+Quick}}{T_{1-Fox}}$$

Note that we only need the first two, as those are the two that tie the coordinate components back to the original force.

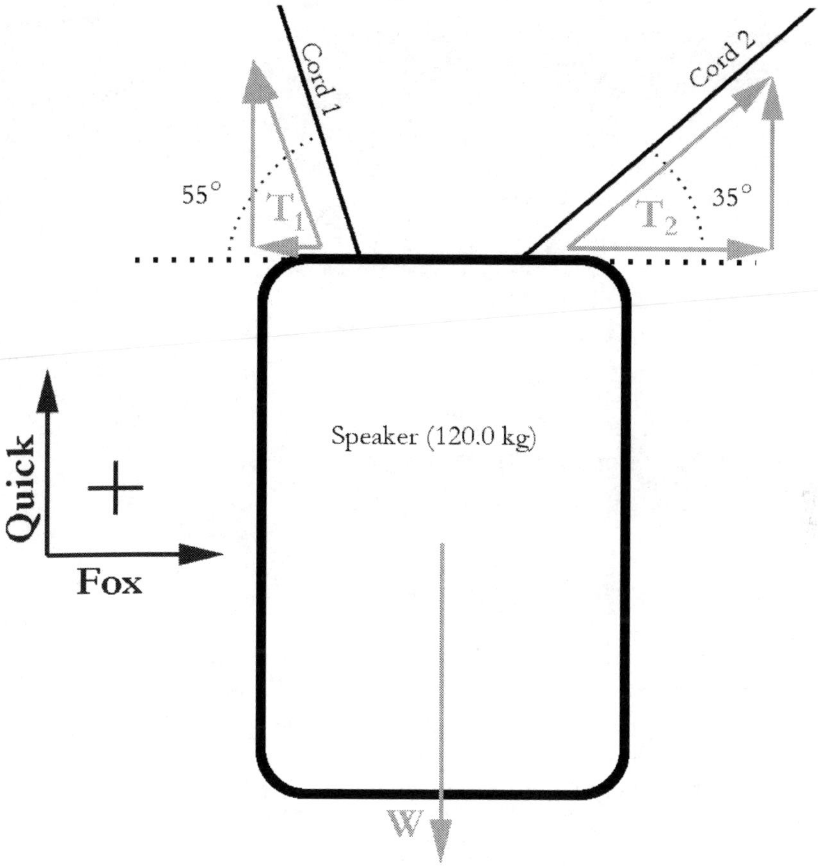

So if we rearrange the first two equations, we'll find that the "Quick" component of T_1 is

$$T_{1+Quick}=T_1\sin(55^o)$$

and the "Fox" component of T_1 is

$$T_{1-Fox}=T_1\cos(55^o)$$

Using SOHCAHTOA for the other cord works similarly, and we'll find that the "Quick" component of T_2 is

$$T_{2+Quick}=T_1\sin(35^o)$$

and the "Fox" component of T_2 is

$$T_{2+Fox}=T_1\cos(35^o)$$

Please make sure you can look back at the diagram on the previous page and reproduce them. (After all, on exams and homework you'll have to do these steps yourself!)

If we add the components of T_1 and T_2 to our little reminder list, we'll have:

"Quick": **–W , +T_1 sin(55°), +T_2 sin(35°)**

"Fox": **–T_1 cos(55°), +T_2 cos(35°)**

Note that I've recorded the signs as well. Any component vector that pointed in the same direction as the coordinate got a positive sign – for example, the "Quick" component of the tension in cord 2 and the "Fox" component of the tension in cord 2. Any component vector that pointed in the opposite direction as the coordinate, such as the "Quick" component of the weight or the "Fox" component of the tension in cord 1, got a negative sign. We'll use this small list in the next step.

6. Apply Newton's second law

$$\vec{a}=\frac{\sum \vec{F}_{ext}}{m}$$

Now that we've broken all of our forces up into our coordinate system directions, we can apply Newton's second law in each direction and build the equations we need to solve the problem. Because acceleration is also a vector, it should also have components in each of our coordinate directions. So instead of fighting our way through vector notation, we'll just write down an equation for each direction:

$$a_{Quick} = \frac{\sum F_{Quick}}{m} \quad \text{and} \quad a_{Fox} = \frac{\sum F_{Fox}}{m}$$

Because we already have a list of the forces for each direction, along with the signs for each, writing down the actual equations is now trivial. Compare the following equations to the little reminder list on the previous page. Hopefully you can see how each term in the little reminder list is part of the sum of the forces for the appropriate direction.

"Quick": $$a_{Quick} = \frac{\sum F_{Quick}}{m} = \frac{-W + T_1 \sin(55^o) + T_2 \sin(35^o)}{m}$$

"Fox": $$a_{Fox} = \frac{\sum F_{Fox}}{m} = \frac{-T_1 \cos(55^o) + T_2 \cos(35^o)}{m}$$

7. Do math and solve

We're almost done with this problem! All we have to do now is solve for the tensions in the cords (T_1 and T_2). The first thing to recognize is that once Indrik hangs the speaker, it should stay in place. That means that not only should it not be moving, it shouldn't *start* moving either. That implies that the accelerations should be zero:

$$a_{Quick} = \frac{\sum F_{Quick}}{m} = \frac{-W + T_1 \sin(55^o) + T_2 \sin(35^o)}{m} = 0$$

$$a_{Fox} = \frac{\sum F_{Fox}}{m} = \frac{-T_1 \cos(55^o) + T_2 \cos(35^o)}{m} = 0$$

Now we've got two equations and two unknowns – we should be good to go! I won't drag you through every step of my calculations, but I will briefly describe the algebraic process. First, solve for T_2 in the second equation, and plug it into the first equation. Now the first equation only has T_1. Solving for T_1 should give you

$$T_1 = \frac{W}{\cos(55^o)[\tan(35^o)+\tan(55^o)]}$$

that you can then use to find T_2, which will be

$$T_2 = \frac{W}{\cos(35^o)[\tan(35^o)+\tan(55^o)]}$$

I highly suggest that you go through the algebra to find these for yourself. We can finally plug in all the numbers ($m = 120.0$ kg and $g = 9.81$ m/s^2) to find the final values,

$$T_1 = 964.3\,N \quad \text{and} \quad T_2 = 675.2\,N$$

Now, you might have calculated the weight of the speaker and found out that it's 1177 N. Based on this, our numbers look wrong at first glance. After all, 964.3 N + 675.2 N is a lot bigger than 1177 N. However, a little bit of thought will clear up this issue. The two tensions don't point in the same direction, so we can't add them up directly. Because the cords aren't vertical, they're not only pulling upwards to hold the speakers, they're also pulling outwards, against each other. Because of this, only part of each tension force is supporting the speaker. The other part – the horizontal part – is canceling with the horizontal part of tension force from the other cord.

Example 2

Joseph Gascho throws a ball at an angle of 30° up from horizontal from the Empire State Building 86th floor open-air promenade at 1.53 meters per second. The 86th floor promenade is 320. meters high. Ignoring air resistance (and the surrounding buildings), how far away from the base of the Empire State Building will the ball hit the ground?

1. Prepare

What happens when someone throws a ball? It falls! The higher up the thrower, the further the ball will go before it hits the ground! This is one of a class of problems called **projectile motion** problems, because all projectiles (like fired bullets, launched arrows, and thrown balls) all behave similarly in the absence of air resistance. In fact, the military now has specially designed computers attached to all of its artillery[46] pieces to do these calculations automagically. However, the basic principles apply just as much to a ball thrown from the observation deck of a building.

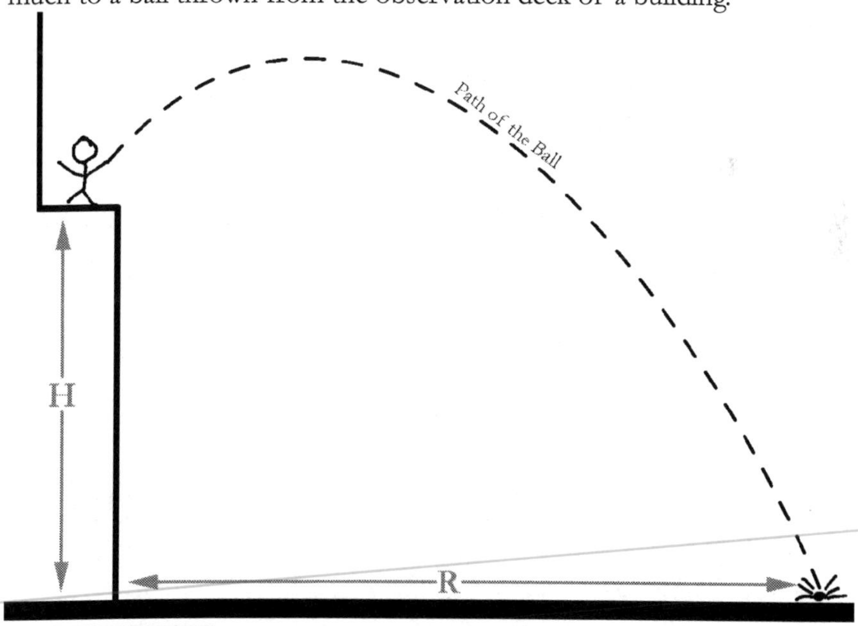

46 **Artillery** (*noun*) – Really big guns that shoot very long distances (long range)

2. Draw a picture

Hopefully this problem suggests an easy and quick picture like the one on the previous page. Notice that although the picture is simple, it captures Joseph, the Empire State Building 86th floor promenade, the falling ball, and where the ball will land. All it needs now is a bag of chips!

It looks great, doesn't it? The only problem is that it's not as useful as we might like. What and where is the object of interest? Although this problem does mention a person and a location by name, when you think about the main action it should become clear that the object of interest (and the only key element of the problem) is the ball. After all, the problem specifically asks about where the ball will land. The previous picture is good in that it gives us an overview of what's going on in the problem, and even allows us to name a few variables. (For example, that drawing lets anyone who looks at it know that R, the distance away from the building that the ball lands, is what we'll eventually be looking for.) A better picture would be the following:

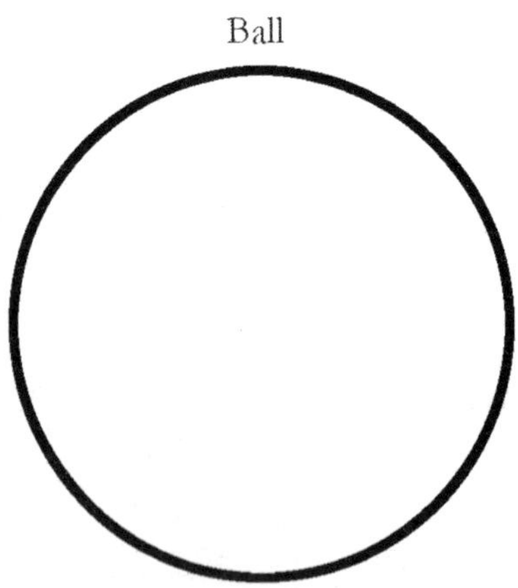

This picture isn't as interesting, but it clearly shows the object of interest, and leaves plenty of room for the addition of forces in the next step. The variables we've named in the previous picture are useful, however, so we'll keep that, too. While we must have one picture (to build our force diagram), there's no rule that says we can't have more!

3. Identify the forces

Looking at the problem statement, what forces do you think will be acting on the ball? We're close to the Earth's surface, so there should certainly be a gravitational force (or weight) acting on the ball. What about some of the others we listed in the previous chapters? Is there a tension? Normal force? Spring force? Electromagnetic force? Wilberforce?

Most of those are silly. There's no cord or string to exert tension, no spring to exert a spring force, and no charge was mentioned, so we don't have to worry about electric force or magnetic force.[47] But what about the ground pressing up against the ball or Joseph Gascho's hand as he throws the ball?

Remember, we are only interested in the forces acting upon the ball during the time described by the problem. We know that Joseph threw the ball; it's stated in the problem. However, because we're given the ball's speed and direction as it leaves his hand, we don't have to worry about whatever force was necessary. The problem starts at the moment the ball leaves his hand, and if the ball and the hand aren't in contact, then there's

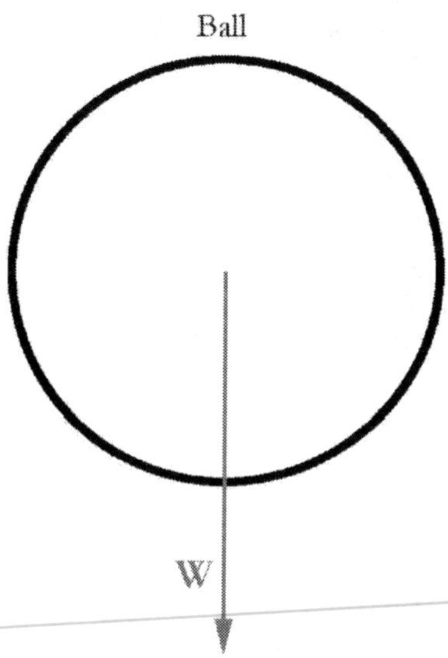

47 And Wilberforce is a university in Ohio. Go Lady Bulldogs!

no force exerted! Similarly, the ground won't exert a force on the ball until they come into contact. At that point, the ball has landed, and we don't care what happens to the ball after it's landed! While the ball is in the air, which is the only time that really concerns us, neither of those forces count. In fact, besides the gravitational field, the only thing that could influence the ball's motion is the air itself, and the problem explicitly tells us to ignore that! The only force acting is the ball's weight, as shown in the picture.

4. Choose a coordinate system

Because there's only one force acting on the ball, this step is trivially easy. We could do something complicated, but let's keep it simple, and point one coordinate axis straight down. The other one has to be at right angles, but it honestly doesn't matter whether it's to the right or to the left – there aren't any forces in either of those directions. However, the ball is heading to the right, so let's choose that direction as positive:

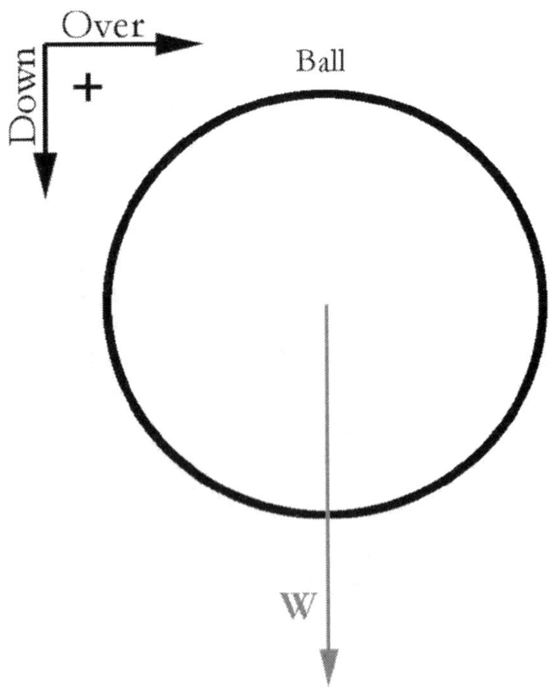

Don't you love it when one of these steps comes out so easily? Don't worry; we'll pay for this simplicity later on (in Step 7).

5. Break forces into components

With only one force and an appropriate choice of coordinates, this step is also less than challenging. The weight points in the positive "Down" direction. If you'd like to see it the same way as the previous example,

"Down": **+W** *"Over"*: *0. Nothing. Nada. Zip.*

6. Apply Newton's second law

Now it's time to translate what we've done into equation form using Newton's second law:

$$a_{down} = \frac{\sum F_{down}}{m} = \frac{W}{m} = g$$

What does this result mean? The ball's motion will be changing only in the "Down" direction; any motion in the "Over" direction will remain constant. (Remember Newton's first law?) So we'll need to break our initial motion into components. That will allow us to change the "Down" piece while leaving the "Over" piece alone.

As an aside, one of the nicest things about projectile motion problems is how easily they resolve, once you figure out what forces apply. As you've seen, this is almost always limited to the weight (gravitational force) of the object. The physics of projectile motion is very straightforward, and that's reflected in the solution we've built using the forces conceptual scaffold.

7. Do math and solve

Of course the "grueling" level of difficulty we've seen so far in this problem had to change at some point. In fact, one of the reasons that I chose this problem was to illustrate that a conceptually easy problem can still be quite challenging because of the mathematics involved.

Let's start with the initial speed. Joseph Gascho threw the ball at an angle of 30° up from horizontal. In the adjacent picture, the angle is marked with the circle segment and the two components to the speed have been added in grey.

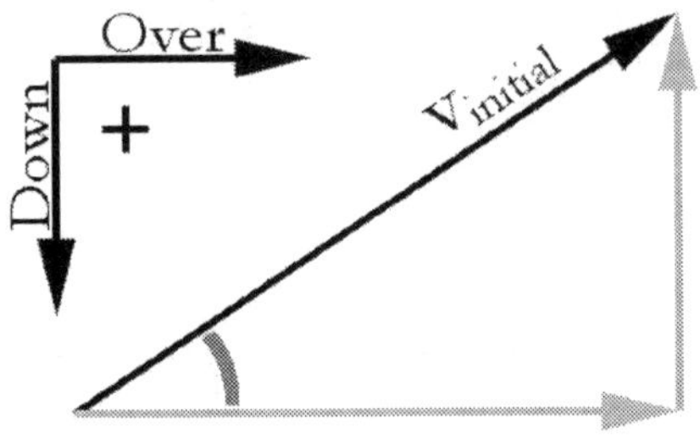

As with problems involving forces, we can use SOHCAHTOA to figure out these components. I've written the pieces below; please make sure that you understand how I got these.

$$v_{down} = -v_{initial} \sin(\theta)$$

$$v_{over} = +v_{initial} \cos(\theta)$$

Unlike the last problem, this problem has a final acceleration. Because of this we'll need the kinematic equations to relate the acceleration to the speed and distance:

$$v_f = v_0 + a \Delta t$$

$$\Delta x = v_0 \Delta t + \frac{1}{2} a (\Delta t)^2$$

$$v_f^2 = v_0^2 + 2a \Delta x$$

Wow, that's a lot of equations! To make sense of all of this information, let's go back to the problem. What are we trying to find? What's necessary to get there?

We want to know how far from the building the ball hits the ground. For that to happen, however, the ball has to hit the ground! In other words, the ball should fall a distance H (the height of the promenade). In the time it takes to do that, it will move horizontally a distance R. That distance, R, is what we want. I've included a copy of the picture below.

This is why we kept this picture! Even though it wasn't useful for the previous steps, it gives us some insight into setting up our equations. Basically, we need two equations – the first will tell us how long it takes for the ball to fall the distance H (in the "Down" direction). The second equation will tell us how far the ball will go in the "Over" direction, given that it only has an amount of time as given by the first equation. In other words, the first equation will tell us how long the ball will be in the air; the second equation will tell us how far the ball gets in that time.

Now we need to choose which equations we want to use. We need something that relates speed, distance, and time. There's only one kinematic equation that fits:

$$\Delta x = v_0 \Delta t + \frac{1}{2} a (\Delta t)^2$$

We can use this for both the "Down" and the "Over" directions. In the "Down" direction, the ball's change in location (Δx) will be the height of the building, H. The initial speed (v_0) will be the speed we found earlier, v_{down}, and the acceleration (a) is what we found in Step 7, g.

$$H = -v_{initial} \sin(\theta) \Delta t + \frac{1}{2} g (\Delta t)^2 \quad (1)$$

In the "Over" direction, the change in location for the ball will be the distance R and the initial speed will be v_{Over}. Because there were no forces in the "Over" direction the acceleration here is zero!

$$R = v_{initial} \cos(\theta) \Delta t + \frac{1}{2}(0)(\Delta t)^2$$

or

$$R = v_{initial} \cos(\theta) \Delta t \quad (2)$$

Now we have two equations, labeled (1) and (2), and two unknowns, Δt and R. However, we only need one of them (R) to answer the question. This presents us with two possible ways of doing the math. The more straightforward method would be to find Δt in equation (1) then plug it into equation (2). The second (and arguably more elegant) method would be to solve equation (2) for the thing we don't want (Δt), then plug it into equation (1) and solve for R. This has the advantage of taking us straight to the value we want. It's impossible, however, to predict which way will be more work, and either way should give you the same answer.

Ultimately, the best choice is the one that makes the most sense to you. Because most people find the first method easier to do, I'll start by solving for Δt from equation (1). If you remember your algebra, you should immediately recognize that we're facing a quadratic equation. Solving those requires remembering the quadratic formula and applying it:

$$\Delta t = \frac{v_{initial} \sin(\theta) \pm \sqrt{v_{initial}^2 \sin^2(\theta) + 2gH}}{g}$$ [48]

48 If you don't understand where this equation came from please review your algebra textbook or "The First Semester Algebra Survival Guide". In no way am I promising or suggesting that I will write "The First Semester Algebra Survival Guide".

The big problem with the quadratic formula is that ± sign that always crops up, representing the two mathematical solutions to the quadratic equation. Generally you can simply grab the positive sign and run with it, which is what we're going to do here. Plug the result above into equation (2) from the previous page and we get

$$R=\frac{v_{initial}\cos(\theta)}{g}\left[v_{initial}\sin(\theta)\pm\sqrt{v_{initial}^{2}\sin^{2}(\theta)+2gH}\right]$$

At this point, we can plug in all of the numbers to find R. From the equation statement $v_{initial} = 1.53$ m/s, $\theta = 30.o$, and H = 320. m[49]. Being careful with the calculator and the math gives a final answer of

R = 10.8 m.

49 The period at the end of the value is there to set the number of significant figures. I haven't talked much about these because they are more of a measurement issue than a physics issue. If you're taking a physics lab you should see a lot more about "sig figs" there. If you're not taking a physics lab, you should be.

Example 3

A block is launched up a 30° ramp with an initial speed of 2.25 m/s. The coefficient of friction between the ramp and the block is 0.100. The block slides up the ramp until it stops, then it slides back down to the bottom. How long does it take for the block to make the round trip, and how fast is it going when it reaches the bottom of the ramp?

1. Prepare

This is the sort of problem that always leaves me scratching my head. Why on Earth would someone launch a block up a ramp, only to see it slide back down? I honestly have no idea. However, physics problems are often full of these kinds of contrived situations, so we'll have to just roll – err, slide – with it.

Just because the problem describes an implausible situation, however, doesn't mean there aren't good things to learn. In particular, note that the problem specifically mentions coefficient of friction. When we identify the forces, we may want to keep in mind that there should be friction in this problem! You may want to go back and read over the description of friction in the first chapter, as those concepts will come into play soon.

To prepare for the next step, let's imagine what the problem is describing. A block is sliding up a ramp. It slows down, stops, then starts sliding back the way it came. Eventually it reaches its original location. The end. (Don't you love a happy ending?)

There are two things to note about this description. First, the block is first moving up the ramp, then it's moving down the ramp. That implies that the direction of the block's motion changes. Why is that important? Hopefully you remember from your quick review[50] of the friction section that the friction force always tries to keep two surfaces from sliding against each other. If the direction of the slide changes, then the direction of the friction force will change as well. If the direction of the friction force changes, then the acceleration must change as well (due to Newton's

50 You did go back and do a quick review, right?

second law). Because we only do problems with constant (i.e., non-changing) accelerations, this therefore implies that we need to break this problem up into two separate phases: the block sliding up the ramp, and then the block sliding down the ramp. Strictly speaking, each phase is a separate force problem! Sneaky, tricksy professor, hiding two force problems in one!

Ahem.

There's another point to be made. While the problem statement implies that the block will ascend the ramp, then descend the ramp, it's quite possible that things might not work out this way. Potentially the block could slide up the ramp, slow down, stop, and then... stay there, stuck in place. If the friction force[51] between the block and the ramp is high enough, the block won't return down the ramp. It will just sit there quietly on the side of the ramp, looking forlornly down at where it used to be as the timer for its return counts toward infinity. In other words, if the block gets stuck, we can't answer the question as posed. The questions asked are meaningless if the block never makes it back to the bottom of the ramp.

None of this, however, affects the way we approach the problem; the basic physics is the same regardless of whether the problem statement is consistent with its solution. Therefore, we're going to take the description given in the problem in good faith and assume it's correct. If it's not, we'll find out as we work[52]. Remember, the physics is in how we construct our solution to the problem, not in the actual numeric answer.

Grab your paper, pencils, pieces, parts – anything you need to help solve the problem[53]. Take a deep breath, hold it for a second, then let it out. Time to git'er done!

---

51 The problem doesn't indicate whether the coefficient of friction listed is for a moving object (kinetic) or a stationary object (static), so we can assume that the value given is for both.

52 Although professors can be sneaky and tricksy, these kinds of problems are almost always honest mistakes – he or she probably didn't intend for the problem to turn out the way it actually did!

53 Coffee, anyone?

2. Draw a picture

Draw a block is traveling on a ramp. That shouldn't be too challenging:

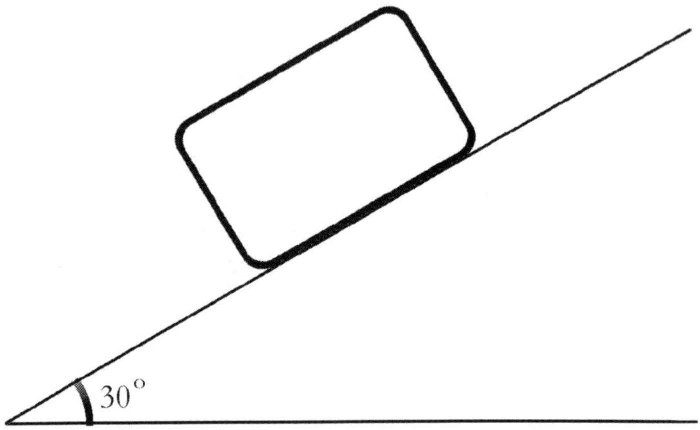

One nice thing about this picture is that it doesn't matter which direction the block is traveling. A block going up a ramp looks a lot like a block going down a ramp. (It also looks a lot like a block just sitting on a ramp, but let's not worry too much about that right now.)

Note that I haven't added any markers showing which direction the block is traveling. Although it might be tempting to add the motion to the picture, please resist. Remember, this is the foundation for our force diagram; adding extra arrows for things that aren't forces will very likely confuse us in later steps.

3. Identify the forces

We already know we need to add a friction force. What other forces should be present? Let's go through the list. Gravity? Check! There's no indication in the problem statement that this doesn't take place near Earth, so we should assume that it does. Normal force? Check! The block sits on top of the ramp, so it's reasonable to assume that the ramp will push upwards on the block to hold it up. Friction? Check! We've already noticed the problem mentions friction; even if it hadn't, however, we have two surfaces sliding (or trying to slide) against each other. In either case, we need to worry about friction. Tension? Nope. The problem didn't involve any ropes or cords. Elastic force? Nope, no springs[54]. I'll leave it to you to explain why the electric force, magnetic force, and thrust don't apply to this problem.

Weight always points straight down, the normal force always points directly out of the surface, and the friction force always opposes the direction of the slide. So we'll need two force diagrams – one for the first phase, when the block is sliding up the ramp, and one for the second phase, when the block is sliding back down the ramp. In the first case, the block is sliding up the ramp. To stop that slide, friction must point in the opposite direction: down the ramp.

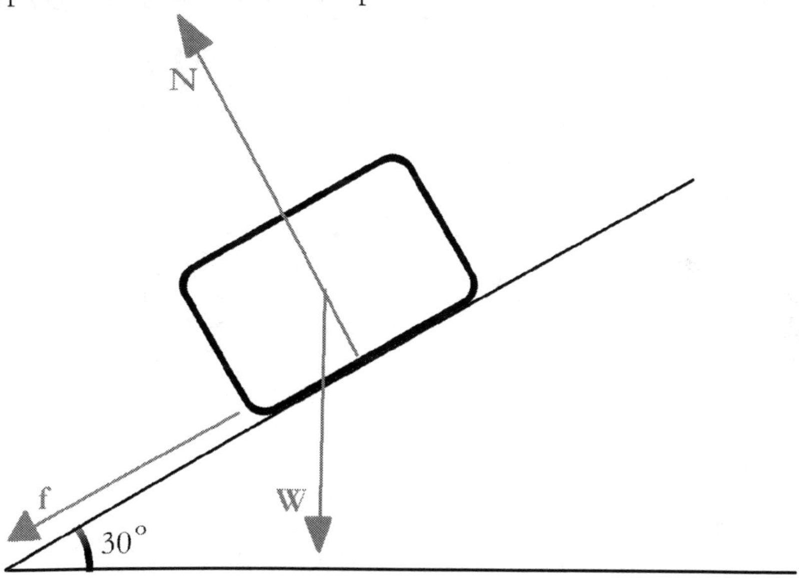

54 "But what if the block was launched up the ramp using a spring?" Ah, great question! Just as in the last problem, however, that would have happened before the problem started. When we come in the block is already moving; it doesn't really matter what happened before that.

For the second phase, the normal force will still point straight out from the ramp, and the gravitational force (weight) of the block will still point toward the center of the Earth (straight down). The only force that will change is the friction force! Because the block is traveling in the opposite direction, the friction force will point in the opposite direction as well.

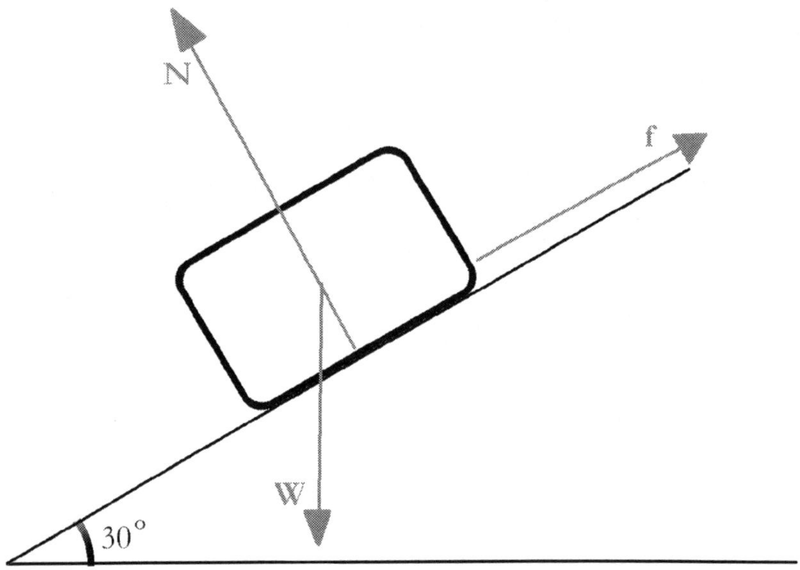

Notice that the only change between these two situations is the direction of the friction force. Perhaps we can use this to our advantage?

4. Choose a coordinate system

As always, we can choose any coordinate system that we want. Something to note, however, is that the block is going to remain on the ramp. The motion of the block is either up the ramp or down the ramp. It would make sense, then, for us to choose one of our axes as pointing in one of those directions – either up the ramp or down the ramp.

There's another reason for this choice. If one axis is parallel to the ramp, then the other must be at right angles – perpendicular to the ramp. This just happens to be the same directions as the friction force (parallel) and the normal force (perpendicular). If we choose a coordinate system that is tilted to match the ramp, then we only have one force, the weight, to break into components. I'm all for anything that minimizes our work!

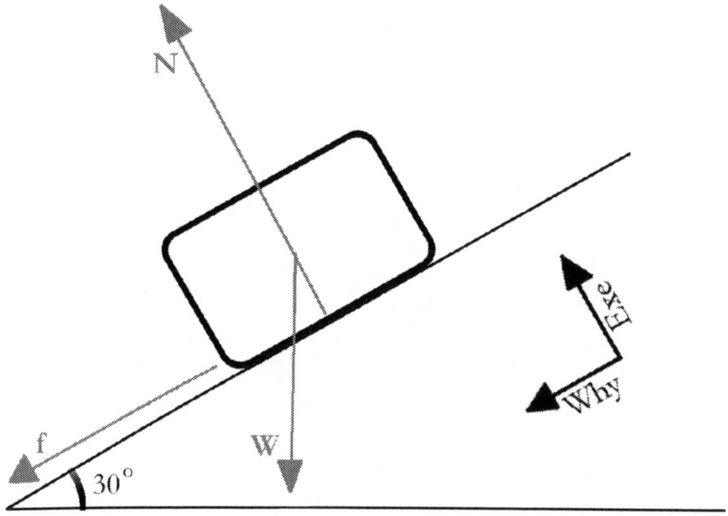

Note that these coordinates work great for the second phase as well; the only change will be that the friction force will become negative when the direction changes. Make a mental note about that. We'll return to it later.

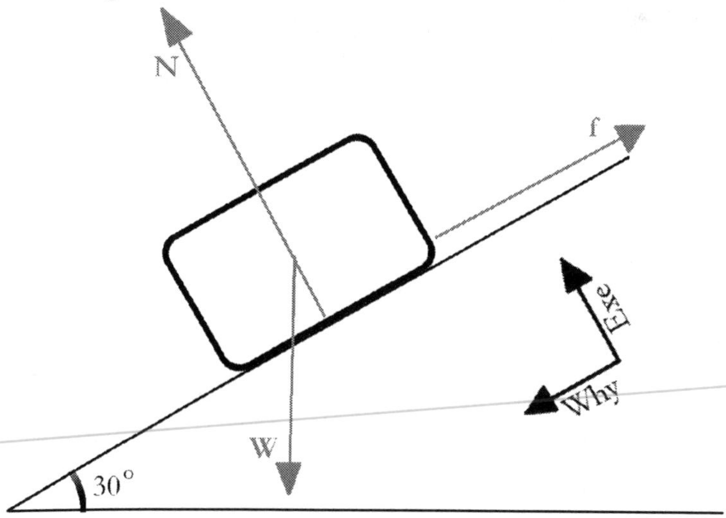

5. Break forces into components

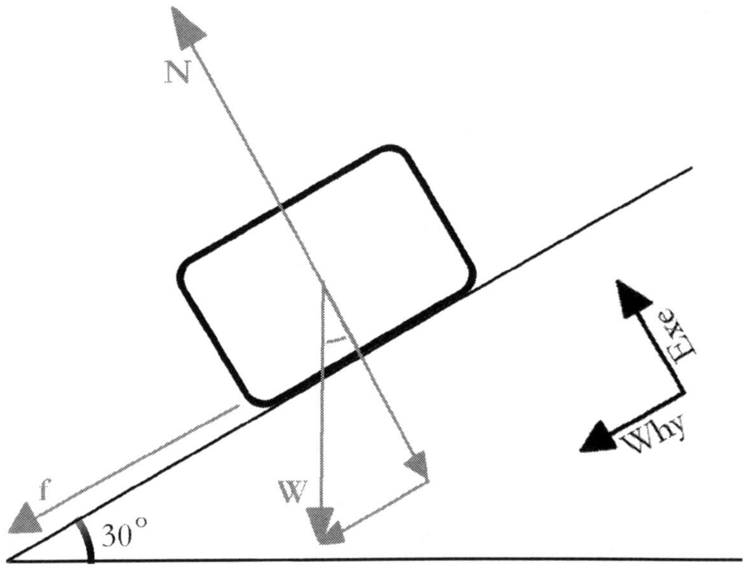

This phase is fairly straightforward; the weight is the only force that needs to be broken into components. You should figure out how to get the components I've included in the reminder list below. I'll give you two hints: First, I've marked the angle in the drawing that I used to build the equations below. You can use geometry to figure out what that angle should be. Second, as always, use SOHCAHTOA! The reminder list (with friction subtitled with the phase) for the first phase of the problem is:

> *'Exe'*: **+*N*, –W cos(30°)**
>
> *'Why'*: +f_1, **+W sin(30°)**

Taking into account everything we've noted about the friction force, we can actually write down a reminder list for the second phase without redoing the problem! The only change will be that the direction of the friction force changes. So for the second phase, we insert a negative sign:

> *'Exe'*: **+*N*, –W cos(30°)**
>
> *'Why'*: **–** f_2**, +W sin(30°)**

6. Apply Newton's second law

At this point, we're ready to put together Newton's second law for each direction. In addition, we've got the information to write the equations for both phases of the problem!

$$a_{Exe}=\frac{\sum F_{Exe}}{m} \quad \text{and} \quad a_{Why}=\frac{\sum F_{Why}}{m}$$

Using the little reminder list on the previous page, we can write the actual equations. Note again that the block should stay in contact with the surface of the ramp. We don't want it to go lifting off into space or to sink down into the ramp itself. In other words, the block shouldn't accelerate in the "Exe" direction, so we can set that acceleration equal to zero:

$$\text{"Exe":} \quad a_{Exe}=\frac{\sum F_{Exe}}{m}=\frac{+N-Wcos(30^o)}{m}=0$$

$$\text{"Why":} \quad a_{Why}=\frac{\sum F_{Why}}{m}=\frac{+f_1+Wsin(30^o)}{m}$$

In the second phase, nothing changes except the sign on f:

$$\text{"Exe":} \quad a_{Exe}=\frac{\sum F_{Exe}}{m}=\frac{+N-Wcos(30^o)}{m}=0$$

$$\text{"Why":} \quad a_{Why}=\frac{\sum F_{Why}}{m}=\frac{-f_2+Wsin(30^o)}{m}$$

These three[55] equations effectively tell us everything we need to know to find the numbers the problem asked for; all that's left to do is the math.

---

55 The "Exe" equation is the same for both phases!

7. Do math and solve

This problem asks us for two numeric answers: the time it takes for the round trip, and the speed of the block as it returns to the bottom.

In order to calculate the time, we need to know how long it takes for the block to come to a stop as it travels up the ramp. We then need to know how long it takes for the block to come back down. Let's start with the time to go up.

First Phase: Block going up the ramp

From the "Exe" equation (from either phase), we find that the normal force is directly related to the weight:

$$+N = W\cos(30^{o})$$

The friction force is always related to the normal force[56], and weight is related to mass and the gravitational field

$$f = \mu N \quad \text{and} \quad W = mg$$

so we can use this and the "Why" equation

$$a_{Why1} = \frac{+f_1 + W\sin(30^{o})}{m}$$

to solve for the acceleration of the block on its way up the ramp

$$a_{Why1} = g\left[\sin(30^{o}) + \mu\cos(30^{o})\right]$$

Once we have the acceleration, which is constant for all of the first phase, we can use the kinematic equations. But which one? The main limitation we have is that the block has to come to a stop. That means that the time interval ends once the block's speed becomes zero.

56 You did go back and do a quick review of friction from the first chapter, *right*?

Looking over the kinematic equations,

$$v_f = v_0 + a\Delta t$$

$$\Delta x = v_0 \Delta t + \frac{1}{2} a (\Delta t)^2$$

$$v_f^2 = v_0^2 + 2\mathrm{a}\,\Delta x$$

The first is the one that fits what we need. It relates the final speed (which for us is zero) to the initial speed (which was given in the problem), the acceleration (which we found above), and the time interval (which is what we want). The only possible tripping point here is that initially the block is traveling up the ramp, in the negative "Why" direction. This means that v_0 is actually negative:

$$v_f = -v_0 + a\,\Delta t = 0$$

Plugging everything in and solving for the time interval gives

$$\Delta t_1 = \frac{v_0}{g\left[\sin(30^o) + \mu\cos(30^o)\right]}$$

Remember, this is for the first phase, hence the subscript on Δt. As it turns out, for the second phase we'll need to know how far the block traveled up the ramp[57]. Using either of the other kinematic equations, we can plug our values in and find that distance:

$$\Delta x_1 = \frac{v_0^2}{2g\left[\sin(30^o) + \mu\cos(30^o)\right]}$$

57 Trust me on this; once you've done enough of these kinds of problems you begin to anticipate these kinds of things.

You may or may not have gotten a negative sign. Because the sign is simply carrying direction, you can drop it from the equation. We only want the distance.

Second Phase: Block coming down the ramp

This phase will proceed very similarly to the previous phase; however, this time our acceleration will come out to be

$$a_{Why2} = g\left[\sin(30^o) - \mu\cos(30^o)\right]$$

As I said, it's possible that the block will never come down, and here you can see why. The part in the brackets has a negative sign! If that part of the equation is zero, then the acceleration for the second phase will also be zero. If the acceleration is zero, then the block's motion won't change. The end of the first phase (and thus the start of the second phase) occurred when the block finally stopped moving. If the acceleration is zero at that point, then the block won't ever start moving again!

So before we do a bunch of work, let's make sure that the piece in the parentheses is greater[58] than zero. We want

$$\sin(30^o) - \mu\cos(30^o) \geq 0 \quad ;$$

for this to be true, then

$$\tan(30^o) \geq \mu$$

must be true. Fortunately, the tan(30°) = 0.557. Whew!

Based upon this, we now know that the block will slide back down the ramp. Let's find out how long it takes and how fast it's going when it's

58 Mathematically this part of the equation could be negative; however, that would imply that the block would accelerate up the ramp. If we look back at the drawing that would imply that the friction force was causing the block to slide more and more, which is directly contrary to the nature of friction. You did go back and review the section on friction in the first chapter, right?

done. Both of these simply require use of the kinematic equations. In particular, we can use

$$\Delta x = v_0 \Delta t + \frac{1}{2} a (\Delta t)^2$$

to find the time and

$$v_f^2 = v_0^2 + 2a \Delta x$$

to find the speed. We know that the distance the block will slide in the second phase is the same as the distance it slid in the first phase; in addition, the initial speed in the second phase is zero, the same as the final speed for the first phase. That simplifies the equations to

$$\Delta x_1 = \frac{1}{2} a_{Why\,2} (\Delta t_2)^2$$

and

$$v_f^2 = 2a_{Why\,2} \Delta x_1$$

Plugging in what we've already found for the distance and the accelerations, and solving for the time interval and speed gives

$$\Delta t_2 = \frac{v_0}{g\sqrt{\sin^2(30^o) - \mu^2 \cos^2(30^o)}}$$

and

$$v_f = v_0 \sqrt{\frac{\sin(30^o) - \mu \cos(30^o)}{\sin(30^o) + \mu \cos(30^o)}}$$

Note: Getting to this point required a lot of careful algebra. Please make sure you can reproduce this equation by working carefully with the equations you've found. Trust me – it's good practice.

The total time is the sum of the time for the two phases, or

$$\Delta t = \Delta t_1 + \Delta t_2$$

or

$$\Delta t = \frac{v_0}{g\left[\sin(30^o) + \mu\cos(30^o)\right]} + \frac{v_0}{g\sqrt{\sin^2(30^o) - \mu^2\cos^2(30^o)}}$$

At this point, we can plug in all of the numbers – v_o = 2.25 m/s, μ = 0.100, and g = 9.81 m/s² – to find our final answers:

Δt = 0.857 s and v_f = 1.89 m/s

Another satisfied ~~customer~~ problem!

Energy

Energy: Conservation

When you think of energy, what comes to mind? Most responses to this question can be put into one of three categories. The first category is what I call "device energy". This is the energy that powers stuff[59], whether mechanical or electrical. Examples from this category include devices that plug into a wall outlet, the power plants and the materials that fuel them, and the infrastructure surrounding and connecting them. It describes the energy of cars, planes, and trucks; of refrigerators, ovens, furnaces and air conditioners; and of televisions, radios, and computers.

The second category is "person energy" – the energy we talk about having or (more often) lacking, regardless of source or use. This category contains the energy that we try to get from sugary soda and coffee in the morning and energy drinks and candy bars in the afternoon. Other examples include the energy of the exuberant child, as well as the lack of energy of the harried parent. It is the energy of social interaction, intellectual engagement, and inherent curiosity. The mystical energies of various traditions also belong here, regardless of whether those energies are channeled through discipline, meditation, exercise, numerology, prayer, psychic prowess, tarot cards, or whatever.

The last category I call "textbook energy", because these descriptions most often sound like someone pulled them out of a textbook or dictionary. However, my favorite description of this category is "Energy is the ability to do work." In fact, I'd bet fictional money[60] that this line, or something like it, appears in your physics textbook.

59 Yes, "stuff" is a technical term.

60 I'm offering fictional money because a) I don't know who you are or where you are to track you down when you lose, and b) because I hate risking real money on bets even more than I hate getting involved in land wars in Asia or going in against a Sicilian when death is on the line.

This, then, is where I will begin, because many courses use this definition as an introduction to energy, and it is wonderfully, terribly wrong.

"Whoa!" I can hear you saying. "Did you just say that the physics book is WRONG?!" As a matter of fact, I did. The issue is more than simply a matter of definition. Look in your textbook for the definition of work. If you find one, it will probably be something along the lines of "a transfer of energy." Well, that wasn't very useful, was it? Define energy in terms of work, then define work in terms of energy. If you don't present those ideas at the same time, you might even get away with it sounding plausible. Unfortunately, almost everyone agrees that this set of definitions is the set to use, and they gloss over the fact that they're talking in circles[61]. And you know why?

Because we, as physicists, have absolutely no idea what energy is. None.

That may seem like a huge admission. After all, if we don't know what one of the most fundamental concepts in our field is, then why study this crap at all? As it turns out, you don't have to know what something is to study it and learn a lot about it. We've already run into a concept with a similar problem: mass[62]. The nature of energy, much like the nature of mass, is a complete mystery. Both are simply "properties of matter,"[63] a description that is infuriatingly nondescript. However, just because we don't know what either one *is* doesn't mean we can't describe what they *do.* That's the approach I will be taking, and to do that, I'll be using an analogy.

Energy ~ Money

To explore this analogy, we're going to have to step outside of our current subject for a bit. Instead of thinking about physics and science and math,

61 **Self-Referential** (*adjective*) – Definition available in the footnotes of page 96.

62 Note that the discovery of the Higgs boson doesn't really explain mass. It just pushes the mystery off onto a new particle like a robber fencing stolen goods.

63 In addition, the two are connected. Albert Einstein's theory of special relativity predicted that one could be changed into the other (and vice versa); millions of experiments conducted since have repeatedly confirmed this.

I want you to think about money and wealth and... well, math. In particular, think about your money – where it comes from, how you use it, and where it goes when it leaves you.

All of the money you have or spend originally came from someone else. It may have been given in exchange for nothing (gifted), given in exchange for a service (earned), given in exchange for a product (paid), taken through trick or force (stolen), or somehow recovered from where it had been lost (found). In all of these cases, however, someone had it before you, and gave, paid, or lost the money you found. Similarly, all of the money that leaves you goes to someone else. It may be gifted, paid to another for a service or product, stolen from you, or lost. Eventually, however, it will end up in the hands of another person. And hopefully it's obvious, but just in case it's not: Any money doesn't leave you stays with you.[64]

For example, let's say someone gives you $100. You treat your friend to a movie with popcorn and drinks, which costs you $25 of that money, then you use another $40 to put gas in your car. Finally, on the way home, you spend $15 on snacks at the grocery store. At the end of the day, you should still have $20 in your wallet.

But what happened to the $25, $40, and $15 you spent at the movie theater, the gas station, and the grocery store? It was used to pay employees, purchase supplies or products, and take care of bills. Perhaps that $25 was used to buy popcorn from a local farmer. Maybe that $40 went to the electric company so that the gas station's lights would stay on. Your $15 could have become part of the store manager's paycheck. Each of those companies/individuals does something with the money as well. Maybe the farmer stuffs the money into an old cigar box hidden in his closet. The electric company could pay taxes to the state. The store manager probably put his money into his bank account. The money in the closet will stay there, trapped in the cigar box. The state will pay a paving company to repair a road. The bank will "loan" the store manager's money to someone wanting to buy a house. And so on, and so on, and so on.

64 You may notice that I have not mentioned counterfeiting new money or burning old bills; I'll address those points in a bit.

Notice that we haven't introduced any new money. That same \$100 that was given to you simply moved on, passed from one entity to the next. Some of it got stuck, like the \$25 in the farmer's closet, but it hasn't vanished. It's still there, waiting to return into circulation. Most of it, however, kept being transferred during interactions. Two individuals came into contact (the store and the manager, for example); as a result of that contact, one gained some amount of money, and the other lost an equal amount of money.

Energy is exactly like this. An object can gain energy (like the \$100 you gained as a gift), lose energy (like the \$40 you paid to the gas station), or simply hold energy (like the \$20 you had left in your pocket at the end of the day). Energy lost by an object is always gained by some other object (the \$40 you lost was gained by the gas station). An object doesn't add energy to the system; it just passes it along or holds onto it. Effectively, energy runs the evolution of the Universe just like money runs the evolution of human economies. This basic concept of energy ~[65] money encapsulates almost everything you need to know about energy.

Originally, money was tied to a material substance – a silver standard or gold standard, for example. That served to fix the total amount of money in circulation. Under this limitation, all of the action in the economy came from how the distribution of money changed over time. Those changes were a product of the interactions that transferred money from one entity to another. Energy is exactly like this. As far as we can tell, there is a fixed amount of energy available in the Universe. All of the interesting things that happen in the Universe are the result of various transfers of energy.

However, there are a few places where this analogy fails. First, consider the fact that there's energy (or money) available to trade at all. Where did this (supposedly fixed) supply originate? With money, the obvious answer is "the government." In the case of the United States, the management of the money supply falls upon the Fed – the Federal Reserve System. As I said, the original production of money was based upon the amount of sil-

65 The "~" sign (a.k.a. **tilde**) is used by physicists (and some mathematicians) to denote equivalence – the fact that two things aren't quite identical, but they are very similar or closely related.

ver and gold available, either mixed into the coins themselves or stored in government vaults[66].

For energy, however, the answer to this question ("Where did the fixed supply come from?") is not so obvious. The structure of the Universe itself acts as a sort of "Energy Fed," enforcing the energy rules. However, the original source of the totality of the energy in the Universe is yet another mystery. Is it energy that was externally supplied when the Universe was created? Or is the current available energy supply simply a debt the Universe holds against itself, like much of the national debt? In the money analogy, the first situation is similar to that of the early United States, with its reserves of gold and silver – the "net worth" of the Universe is positive because of external resources, and this extra "wealth" provides the energy available for interactions. The second situation is more like the modern money supply – governments create debt (negative money) to provide cash (positive money) for the economy. In this case the "net worth" of the Universe is zero (or even negative), and the only energy available for use is energy borrowed against the Universe itself. Currently I do not know if there is a way to tell the difference between these situations, as we do not have another Universe with which to make a comparison.

Another flaw of this analogy is the existence of counterfeit money. There has always been a desire to have more money than what is immediately available, and some people have even resorted to creating "new" money through fraudulent means. Forgetting for a moment that this is illegal, there's no corresponding analogy for energy. There does not seem to be any way for an object to simply create energy out of nothing. In the same way, while it is possible to destroy money (for example, by burning the paper it's made of), the Universe does not seem to allow this possibility either. In fact, if we do ever find a way to make "counterfeit" energy[67] or to permanently destroy energy, the implications would be enormous. Such

66 This is extremely simplified, by the way. I am ignoring here much of the complex interaction between different existing currencies available during the founding of the United States, as well as the variance over time of conversion rates between the dollar, gold weight, and silver weight.

67 As an aside, any device which produces "counterfeit" energy is effectively a perpetual motion machine, and vice versa.

processes would allow us to arbitrarily and significantly increase or decrease the overall "net worth" (as mentioned above) of the entire Universe. The consequences of such a momentous change (in either direction) would be quite difficult to predict. Fortunately, the Universe can and does enforce its energy policy directly upon each and every energy interaction. The closest the Federal Reserve can come to this level of enforcement is to encourage individuals to test and check the money supply at each individual transaction[68].

There is, however, one exception to the above. Curiously, under certain circumstances, it *is* possible to create a small amount of mass or energy from nothing. However, the Universe sets a very strict, very short time limit on this. After that time limit expires, the mass or energy *must* be destroyed. If it cannot be destroyed, an equivalent amount of energy is simply *taken* from one of the objects that was involved. This kind of interaction does have a place within our money analogy – it's similar to borrowing money from a loan shark. You won't get much, what you do get you won't have for very long, and you *will* pay it back, one way or another.[69]. This situation does not come up in everyday situations. The Universe only loans out small amounts of energy – barely enough to create an atom or two, much less a car, a baseball, or the head of a pin big enough for an angel prom. Don't worry about it, though – this case won't come up unless and until you take quantum mechanics. The only reason I'm bringing it up is because it's pretty cool.

Anyway, the upshot of all this is that while energy may change hands and change forms, it cannot be created or destroyed. This principle, called **conservation of energy**, is so fundamental (and so useful) that it bears repeating:

Energy can be neither created nor destroyed.

68 For example, one can test the validity of bills using special markers or by looking for special features that are hard to reproduce: raised print, watermarks, special materials, fine details, etc.

69 I am in no way comparing the Federal Reserve to loan sharks. Please don't arrest me, break my knees, or introduce me to the fishes.

Implementations of this idea show up, not just in every nook and cranny of physics, but throughout all of the sciences. It is the root of chemistry, and in many ways forms the foundation of all of engineering. Conservation of energy is perhaps the most fundamental understanding we have of how the Universe works. This is ironic, considering that we don't really know what energy is. However, it doesn't matter, because we know the rules it follows!

What about so-called "energy production"?

But what about those other two categories I mentioned at the start of the chapter? When we talk about device energy, we're often talking about the different ways of "producing" energy: coal, natural gas, petroleum, nuclear, hydroelectric, wind, geothermal, solar, etc. Aren't there are a whole lot of ways to make energy? And there are efforts to "conserve" energy – turning off lights, walking or biking instead of driving, getting energy-efficient appliances, and so on. Surely it must be possible to use up the energy we have if we're in danger of not having enough?

The answer to both questions is "No." There are a lot of ways to produce *useable* energy (either electrical energy, which is used by our refrigerators and air conditioners and computers and light bulbs; or energy of motion, which is used to get things from one location to another). However, the energy was already there – as chemical energy in the case of coal, natural gas, and petroleum; as mass in the case of nuclear; as gravitational field energy in the case of hydroelectric; within the motion of the atmosphere in the case of wind; and as electromagnetic field energy in the case of solar. Power plants that tap into each of those energy sources are simply converting the prior forms of energy into something we can use. That's why a brick of coal is consumed in the power plant – the coal pays out some energy, and the power plant gains some energy. Some of the energy comes out as heat. That energy gets lost, like the pennies that slip out of your hand at the drive-thru window. Neither the heat energy nor the pennies have vanished. In both cases it's just not worth the effort to go after what's been lost[70].

70 This description is a gross simplification of how a power plant actually works, by the way. Actually, most power plant designs actually have an intermediate step that converts almost all of the energy to "heat". Some of that heat en-

Note that, like money, energy comes in different forms. With money, we can talk generally about bills versus coins. We can also get more specific, and talk about $100 bills versus $10 bills versus $1 bills, or about quarters versus dimes versus nickels versus pennies. All of these, however, are specific representations of the same thing – money. In the same way, energy comes in three broad forms (kinetic energies, internal energies, and potential energies), each of which can be broken down into more specific types. All of them, however, represent the same thing. A rose is a rose is a rose, and energy is energy is energy is energy.

Forces and Energy

Before we get into the different forms and types of energy, however, I want to take a moment to look back at forces. When we discussed forces we talked about how a force could change the motion of an object. One form of energy is motion energy. Forces often change an object's motion. Therefore, forces can change an object's motion energy. Based upon what we've discussed here, however, the force can't create new energy in that object, nor can it destroy the energy the object has. Logically, then, the force must have either moved energy into or out of this object from somewhere (or something) else. Quite literally, **forces transfer energy**. Remember back when I said that all forces require contact? This is why I stressed that point. Just as you can't hand over money to someone who isn't within your reach, forces can't transfer energy between objects that aren't touching. You can, however, use an intermediary in both cases. For money, you can transfer money to a bank, and then have the bank transfer money to someone far away. In the same way, an object can transfer energy to a field, and then have the field transfer the energy to a faraway object. Sometimes, however, these transfers aren't as efficient as you might hope. Maybe you decide to give cash to a friend and have the friend carry the money. It would be unfortunate if your friend paid all of his expenses, including gifts and tips, out of your funds before delivering it! You can bet that it will be difficult to get all of that money back once it has been spent. In the same way, sometimes energy can be converted into a form where it is difficult (if not impossible) to recoup, usually because the energy has been spread out in a disorganized or chaotic manner. When a

---

ergy is then converted to electricity, and the rest is lost to the environment. The underlying comparison with the pennies still holds, however.

force delivers the energy in a neat and recoverable manner, that force is called a **conservative force**. Any force that pays its expenses, gives out gifts, and spreads the wealth such that the energy it transferred is not recoverable is called a **non-conservative force** (among other, less polite, names[71]). Friction is the most commonly invoked non-conservative force, but a tricksy[72] professor might invoke a non-conservative applied force of some kind, just to be mean.

For Calculus-Based Physics:

Drag, like friction, is also a non-conservative force. From here on out, whenever I talk about the consequences of friction, you can just as easily apply it to drag as well. Remember, drag is just friction in a liquid or gas.

Let's go back to the friend example for a moment – the one who pays all of his expenses and gives extravagant gifts using the money you asked him to deliver. Suppose you had to send the cash across the country and you didn't have any alternatives. Would it be better to send him on foot or by plane?

Personally, I'd choose the quickest route possible. Hopefully I'll lose less money if I minimize the number of opportunities he has to spend! If he goes by foot, he'll interact with a lot more people; because of that, I can expect that he'll waste a lot more cash.

In physics, this idea is called **path dependence**. The amount of money your friend loses depends upon how he gets to the destination. In the same way, forces that "lose" energy – non-conservative forces – lose more energy when they take the long way. More opportunities for interactions means that more energy is spread out and ends up unrecoverable. By contrast, if your friend is trustworthy, it wouldn't matter how he got to his destination. Conservative forces are path-independent. All that matters in

71 I would propose "moocher force" or "grifter force". I doubt that either of those would really catch on, though.

72 Yes, "tricksy" is a real word. Just ask Gollum. "Sneaky little professors. Wicked, tricksy, false!" **Note:** I am not trying to imply in any way that professors are hobbits or any other kind of fantastic or imaginary creature. Well, maybe Uruk-hai... JUST KIDDING! Please don't sue me.

those cases is where you start and where you end. It doesn't matter how you got from one to the other.

Common Energies in Introductory Physics

Kinetic Energy

Kinetic energies are energies of motion, and are based entirely on the speed of the object, regardless of direction. This is actually one of the great weaknesses of using conservation of energy to solve problems. Conservation of energy may tell you how fast an object is going, but it will not tell you what direction it's traveling. As we'll see later on, this can sometimes spell trouble.

There are two types of kinetic energy. **Linear** kinetic energy is the energy of an object traveling in a straight line. A car traveling down a straight road has predominantly linear kinetic energy. **Rotational** kinetic energy is the energy of an object that is turning about an axis. A merry-go-round has rotational kinetic energy, as do the kids riding on it. A bike tire has both linear and rotational kinetic energy as it rolls along the street. (For now, however, we're only going to deal with linear kinetic energy.)

Calculating kinetic energy is pretty straightforward. As you might expect, it will depend upon the mass and the velocity (motion) of the object:

$$KE_{linear} = \frac{1}{2} m v^2$$ [73]

Potential energy

Potential energies, on the other hand, are a little bit trickier to define. The standard definition is that these are "stored" energies, or are energies "available to do work." We've already discussed the problems with using work in our definitions, so I'm not going to. Another definition is based upon the force that transfers the energy – if the force is conservative,

73 Sometimes kinetic energy is represented in equations by "T"; personally I find KE to be easier to read, so that's what I'll be using throughout this book.

then the energy associated with it is considered a potential energy. While this definition is better, it's still problematic. Instead, I am going to define potential energy as any energy that is "stored" in a field. This creates a nice dichotomy: kinetic energies are associated with objects; potential energies are associated with fields. In addition, using this definition will result in the same set of potential energies as the "conservative forces" definition, with one huge exception – the energy associated with a spring. Most physics books consider elastic energy to be potential energy; under my definition, elastic energy is an internal energy[74]. I'll say more about that when I discuss elastic internal energy.

One thing that potential energies all have in common is that they depend strongly on the arrangement of objects within the field – where is the object in the field, relative to other objects that are influencing the field? Potential energy depends on where the object is; kinetic energy depends on how the object is moving.

Generally, every fundamental field in physics has an energy associated with it. The one that will come up most often in the first semester of physics is the **gravitational** potential energy:

$$PE_{gravitational} = m\,g\,h$$

Let's go over this step by step. Obviously PE stands for potential energy[75]; m symbolizes our object of interest – the object that will either gain energy from or lose energy to the gravitational field due to a change in position. Of course, g represents the value of the gravitational field at the object's position. As a potential field, we need to know the "where": h measures the location of the object – how far the object is above the ground. Note here, like in the section on Forces, we've assumed that the problem, whatever it is, takes place near the surface of the Earth, where the gravitational field is relatively uniform in strength as well as direction.

74 The definition itself doesn't matter so much to the solving of problems – remember, energy is energy is energy is energy. However, you may want to keep the distinction in mind when discussing elastic energy with your professor.

75 Potential energy is traditionally represented by "U"; however, for clarity and consistency (with KE) I will be using PE.

The electric field also has a potential energy associated with it – the aptly named **electric** potential energy. Although it rarely comes up in the first semester of physics, for comparison's sake I'll add one version of it here:

$$PE_{electric} = q\,E\,d$$

Note the similarities to the gravitational field. q represents the charge of interest – the object that will either gain energy from or lose energy to the electric field due to a change in position. E symbolizes the value of the electric field at the charge's height, and that – the charge's height – is represented by d[76]. Just like the gravitational field, this equation assumes that the electric field doesn't change significantly in strength or direction. As it turns out, that assumption is much less useful than it was in the gravitational case – for gravity, we have this gigantic mass[77] that influences everything around us! However, there's no corresponding hunk of charge hanging around to influence the electric field. Because of this there are fewer circumstances in which it is obvious that the electric field will be uniform in both size and direction. Fortunately, you probably won't have to deal with electric fields until your second semester of physics. There are other fundamental fields as well, such as the strong field or the weak field; although these also have associated potential energies, they're much more complex in form. I could tell you more, but then I'd have to kill you[78].

Internal Energy

Internal energies are a third major type of energy and are a bit more abstract than kinetic or potential energy. For example, consider a car. You could talk about its kinetic energy as it drives along the highway, or its potential energy as it rolls over a bridge. However, in both cases you're talking about it as if it were a single object. As any mechanic can attest, it's not. A car contains hundreds of individual components, most of which

76 d is chosen instead of h mainly because the electric field doesn't have to point straight down. Instead, "d" increases in the opposite direction the electric field points.

77 If you can't figure out what this might be, here's a hint: what's directly below you and approximately 12,750km across?

78 I'm just kidding; the nuclear energy stored in the strong field or the radioactivity mediated by the weak field would do it instead. </joke>

can be further disassembled into smaller pieces. In fact, a single compact automobile is quite literally made of thousands, if not tens of thousands, of discrete parts. And even though each of these parts generally moves with the car as it moves down the road and over the bridge, the parts themselves might be moving with respect to the car. Wheels turn, shafts spin, wires rattle, coins slide, people bounce. What goes on inside that car is quite complex!

For the most part, however, we can ignore all of that. As long as that internal stuff remains that – internal – it won't affect anything outside of the car. From our point of view on the side of the road, the rattling of wires and the sliding of coins doesn't affect the car's overall motion. It all cancels out, leaving what we started with – the kinetic and potential energies of the car as a single object as it drives along that highway bridge.

But what if it doesn't cancel out? A car sitting at a red light can accelerate when the light turns green, suddenly gaining kinetic energy from seemingly nowhere. We know, of course, where the energy comes from. The car's engine consumes gasoline, converting the chemical energy in the gas to the rotational kinetic energy of the wheels[79]. The wheels then push against the road to convert that rotational kinetic energy into the car's linear kinetic energy. The gasoline was internal to the car, but it could affect the external motion of the car. Internal energy, then, is any energy that is contained (or hidden) within an object, but can be used to change the kinetic or potential energy of the whole object.

In almost every case, internal energies reduce to either a kinetic or potential energy. However, they almost always involve interactions between a huge number of objects – the atoms and molecules that actually form the structure. Now we're not talking about hundreds of thousands, or even millions; we're talking about trillions of trillions of objects, each of which has at the very least linear kinetic, rotational kinetic, and electric potential energy. Dealing with each and every individual atom is unrealistic, so instead we add up all that hidden energy, and anything that can affect the whole object we call an internal energy.

79 There's some heat released as part of this conversion as well, but we'll ignore that for now.

Let's return to gasoline as an example. Its internal energy is stored in the chemical bonds of the hydrocarbon molecules that make up the gasoline. The chemical bonds themselves are a product of the electric field interactions between the carbon and hydrogen atoms that make up the hydrocarbons. Ultimately, then, the chemical energy of gasoline is really electric potential energy! It's much easier, however, to simply work with the overall chemical energy of the gasoline as a whole than it is to try to work with the electric potential energy of a hundred million billion billion individual carbon and hydrogen atoms[80]. Nuclear energy is also an internal energy. In this case, however, the energy is stored within and between the protons and neutrons that make up atomic nuclei. The energy released in a nuclear reactor or atomic bomb is generated by a huge number of atoms – a hundred billion billion to a few ten thousand billion billion atoms! Because chemical and nuclear reactions usually[81] require so many interactions, it is usually much easier to simply abstract them as "internal" energies set by the amount and type of material.

Just in case you ended up skimming, here's the TL;DR[82] version: Internal energies abstract away the overwhelming internal complexity, and instead represent in a simple way the energy stored within an object can be used by the object.

Another excellent example of this is the **elastic** internal energy. Think of a rubber band. You can stretch a rubber band slowly, then let it go; in this case, the energy stored within it comes out explosively, flinging the rubber band across the room like a bullet. Or you can attach a rubber band to a toy plane's propeller and twist it; upon release, the rubber band will slowly untwist, turning the propeller to move the plane. In both cases, the energy stored within the rubber band itself is released to move the whole rubber band (or another object).

---

80 I mean this literally! A single mole of a substance is about 10^{23} atoms, or a hundred thousand billion billion atoms. A single gallon of gasoline is the equivalent of about a thousand moles!

81 It is possible to use conservation of energy when dealing with a single atom or molecule; in these cases it usually isn't necessary to use internal energy.

82 Stands for "too long; didn't read". Feel free to use it in online forums and email to take the sails out of some long winded blowhard... like me.

Structurally speaking, elastic internal energy is due to bonds between and within the molecules of the material. As the object is stretched, squished, or flexed from side to side, the molecules and atoms within the object are shifted as well. Positive and negative charges jostle and tug at each other, and the bonds try to return them to their original positions. It is this struggle between the movement of the atoms and the bonds forcing them back into place to return that makes the object elastic[83]. Thankfully, we can ignore all of the tortured dancing of atoms and complex gyrations of molecules, and simply describe how the entire object behaves using the elastic internal energy.

In the simplest case, the elastic potential energy depends on how the object is deformed. These deformations can be linear, like the stretched rubber band that is then fired like a bullet, or they can be rotational, like the rubber band that drove the propeller of the toy plane. For now we'll only concern ourselves with the linear case.

$$IE_{elastic} = \frac{1}{2} k \cdot (x - x_0)^2$$

Hopefully it's obvious that "IE" stands for internal energy[84]. The stuff in the parentheses, $(x - x_0)$, is the distance the object is compressed or elongated. In other words, it measures how much it has been squished or stretched. The k is the **spring constant**, and depends on a whole bunch of factors (like the type, number, and strength of those aforementioned bonds). Unless someone gives it to you (or you're solving for it), there is no reason on Venus that you should know what it is.

Note that your textbook – and your prof – will probably refer to elastic energy as a potential energy instead of internal. Like potential energies, the elastic energy depends very much on position. In addition, like field forces, the elastic force is conservative, unlike the forces associated with other internal energies. Strictly speaking, however, elastic energy has more

83 When the bonds actually break, the object's behavior changes. The rubber band isn't as stretchy anymore, for example. If enough bonds break so does the rubber band!

84 Curiously, internal energy is also traditionally represented by "U", just like potential energy; however, for clarity and consistency (with KE and PE) I will be using IE.

in common with heat and chemical energy than it does with gravitational or electric energy. It doesn't matter how you classify it, though – remember: energy is energy is energy is energy.

There is one other really important internal energy. However; because of its importance, I'm going to devote the next chapter specifically to it.

Other Energies and Work-Energy

There are other forms of energy we could discuss – sound and light are two examples that spring to mind – but most of them don't tend to show up much in first semester physics. If they do, simply give them a label and move on. I've said it before, and I'll say it again: energy is energy is energy is energy; just because your professor introduces you to frozbah[85] energy early doesn't mean you treat it any differently than a potential energy or kinetic energy.

What if, however, your professor wants to know how much energy a given force transferred? Yes, that's a fair question; about eight pages ago, I mentioned that forces transfer energy. (If you don't remember, I'll give you a chance to go back and look.) So how do we figure out the energy transferred by a given force?

Let's think this through logically. Forces act upon objects, so if a force is transferring energy, then it must be changing an object's energy. Kinetic energy is associated with objects, so forces must change objects' kinetic energies!

If a force is changing an object's kinetic energy, then the object is either speeding up or slowing down. Remember from our discussion of forces, however, that these changes cannot happen instantaneously. All changes in physics are continuous – they take time. But we're dealing with changes of motion as well. Any change that takes time will also require some distance over which to act.

85 *hand wave* I did not actually mention frozbah energy. Also, these aren't the droids you're looking for. Move along. Move along.

Remember that hypothetical car that hit the deer (back in Chapter 1), then got repaired (in Chapter 2)? Imagine you're driving down that same highway, and another deer walks out onto the road in front of you.

I'll give you a moment to get past the "Oh, no! Not again!" reaction. Once you're done, consider your first reaction. If it's not "slam on the brakes!" - I'll wait. Okay, is it "slam on the brakes"? If not, back up two sentences. If it is, I want you to imagine what happens next. Anything that isn't pinned down or otherwise attached to the car goes flying forward[86] as the car squeals to a stop. Whew, you missed the deer this time. But was it because your car stopped instantly? When you pressed the brake and applied a force to stop your car, how far did the car slide? As your car slowed to a stop (over time), it was still moving forward. The distance it covered in that time period was the distance over which the braking force had to act in order to transfer out all of your car's kinetic energy.

It turns out that to calculate the energy transferred by a force, we need to know this distance over which the force acts:

$$W_{force} = x F \cos(\theta) = \vec{x} \cdot \vec{F}$$

W is the energy transferred by the force, or the work done by the force. It's okay if you don't remember the definition of work I gave back at the beginning of the chapter; a more complete definition is that work is the transfer of energy that results from a force's attempt[87] to change an object's kinetic energy. F is the size of the force. x is the distance over which the force acts, and θ measures the angle between the direction of the force and the direction the object moved. Because this equation relates the kinetic energy of an object to the work done by a force, it is called the **work-energy theorem**.

---

86 ...or, more appropriately, continues moving at the same speed in the same direction until acted upon by an outside force. If you added this statement (or something similar) before even reading this footnote, you go, girl! If you're male, then you go, boy! If you don't consider yourself a girl or a boy, then... You know what? I'll just stick with a gender neutral "Congratulations! You've exhibited physicist thinking!"

87 The attempt may fail. I'll say more about that later.

If you're familiar with vector multiplication, the second part of the equation is the dot (or inner) product version. If you're not familiar with it, don't worry – all it means is that the distance only matters if it's in the same direction of the force. In other words, you only transfer energy if you're pushing in the same direction (or the opposite direction) as the object moves. That's effectively what the cosine means as well. If the force and the direction of the object's shift are the same, then θ is zero, and the cosine of zero is one. If they are at right angles, then the angle is 90°, and the cosine of 90° is zero.

To give you an example, imagine your car has broken down in the middle of the road, and a friend comes to help you push it to the shoulder. Just as you start pushing the rear bumper, he jumps up on the roof and begins pushing downward with all of his might. If the car starts moving, who transferred the energy that's creating that motion? I'm guessing your answer doesn't involve your friend. If the force exerted is not parallel to the object's path, then the force won't change the object's energy. It can't.

For Calculus-Based Physics:

For you calculus folks, the equation is almost identical. However, because you know calculus (lucky you!), you can follow a force as it pushes along a path of whatever shape! For you, the work-energy theorem will look more like:

$$W_{force} = \int \vec{F} \cdot \vec{dx}$$

where dx is an infinitesimal segment of a path. The integral, as always, simply represents the sum over all of the infinitesimal segments that make up the entire path. Please remember that F may change along the path, so it can't always be pulled out of that integral as a constant!

Note that this implies that any force that is always at right angles to an object's motion can *never* change the object's energy. In other words, such a force will never be able to do work. Under most circumstances, neither the magnetic force nor the normal force can change an object's kinetic energy[88].

88 The exception to this for the normal force is almost never discussed, but it comes up during collisions. The exception for the magnetic force shows up in

Next stop: heat! All aboard!

---

the rotation of dipoles and requires the rotational form of the work-energy theorem. The rotational form of the work-energy theorem will be played by Sir "Not Appearing in This Film".

Energy: Heat

So far we've talked about the most basic forms of energy: kinetic energy, potential energy, and internal energy. Although heat fits nicely into the last of those types, it is complex enough to merit its own chapter. Why? Read on, MacDuff!

When Kinetic Energy Becomes Complex

Let's start with a quick review. Remember that internal energies abstract away overwhelming internal complexity. In other words, although we could describe the energy of an object in terms of the individual molecules and atoms, it's both easier and faster to hide all of that complexity behind the facade of an aggregate quantity – the internal energy. When we are dealing with heat, however, we will want to jump back and forth between these two views. To really "get" heat, we'll need to understand what's going on at the microscopic level; to do work with it, however, we'll need to jump out to the macroscopic level, where heat is a simple internal energy of the object.

At the level of individual molecules and atoms, there is no such thing as heat. Instead, each of those small building blocks has kinetic energy – they're flying back and forth, spinning and turning, slamming into each other and bouncing back like headbangers in a mosh pit. It's the sum of this kinetic energy, added up through the entire material that constitutes heat. Almost, that is.

Remember, heat constitutes only the internal energy of the object. If this object is also moving, then it has some overall *kinetic* energy as well. But if the object is moving, however, then all of its molecules are moving as well. Therefore, we have to subtract out the overall motion of the object before we add up the kinetic energy of all of the molecules and atoms. Heat, therefore, is the total kinetic energy within the particles of the object after we've subtracted out the bulk motion of the object. Said another way, it's the random motion of the molecules and atoms that is present even when the object is sitting still.

It's this last part that makes heat problematic at times. It's easy to convert the overall kinetic energy of an object to another form. However, heat, by definition, doesn't include any of that energy. It's the disorganized and random motion that's left over after you've subtracted out all of the organized motion. The difference between these is like that between dogs and cats – dogs that have been harnessed to pull a sled versus cats that are being sprayed by a garden hose. One of these sets will be a lot more productive, work wise, than the other. In other words, once energy gets converted to heat, you might as well be trying to herd angry wet cats. Organizing the motion of the molecules enough to convert that kinetic energy into a different form of energy is basically impossible. Once energy gets converted to heat, it pretty much stays there[89].

Remember our discussion about conservative versus non-conservative forces? Heat is quite often the byproduct of a non-conservative force. I won't name any names[90], but some forces are particularly bad about converting useful energy into heat. This is the equivalent of throwing away perfectly good energy. You wouldn't intentionally toss money out the window to scatter to the wind, right[91]? That's why most engineers try to minimize the effect of these forces in the mechanical systems they design. Non-conservative processes can't be eliminated completely, but the more their effects are reduced, the more efficiently the system will operate.

Of course, sometimes heat does exactly what we want. Cooking, for example, is the application of heat to a substance in order to induce chemical changes that will make the substance more palatable to eat. We manage heat to make our environment more comfortable, and to manipulate the

89 Blackbody radiation is beyond the scope of this book, but it is one way in which heat energy can become a different form of energy.

90 *cough*friction*cough*

91 If you would, please let me know where you live? Um, for no reason in particular, of course.

physical properties of materials during processing. Because of this, I'm going to concentrate on two basic areas of heat study: methods of transferring heat energy, and the basic physical effects of heat changes – temperature changes, phase changes, and size changes.

Heat Transfers

Remember that heat is a form of energy, which means that it can and often will get passed around. As those little molecules and atoms crash into each other and bounce back, they'll slow down and speed up, passing around kinetic energy like chicken pox in a class of kindergarteners.

However, there are limitations on how the molecules and atoms can move, which also limits how they transfer energy. Imagine, for example, if we put each atom in a tiny box, so that it can't collide with the others. Because its interactions are limited, it won't be able to transfer any kinetic energy. Returning to the analogy of the kindergartners, the one that stays home can't interact with the others and won't be able to spread the virus. However, it's hard to find tiny boxes to confine atoms like this. More commonly, the atoms will end up trapped by both proximity and weak bonding with neighboring atoms. In this case, an atom's interactions are limited to other atoms that are nearby. Macroscopically, this state is described as a solid. In order for heat to cross such an object, the energy has to be passed across the material consecutively, from one atom to the next. This process is called **conduction**. This is the case of the kindergarteners mentioned before. Those who are in the same class come into close contact, and the virus spreads directly from one child to each of his or her neighbors until it has spread among the entire class.

But what if one of those children walks out of the classroom, carrying the virus to another room? Atoms can do this as well. If the atoms lack some or all of the weak bonds, they are less confined. They're no longer as limited in interacting only with neighbors, and can move further distances, carrying the energy with them. When the motion of the atoms and molecules themselves becomes the dominant mode of energy transfer, we say they are in a state of **convection**. The energy is carried with the motion of the atoms. Because the atoms must be able to move to do this, convection occurs almost exclusively in liquids and gases.

The third type of heat transfer is less interesting to us, but to be thorough I'll mention it anyway. Imagine that one of the sick kids in the hallway sneezes toward the end of the hallway[92]. Even if the child doesn't walk to the far end of the hallway, the virus will make it down there, carried by the sneeze. In fact, a child at the end of the hallway can easily end up sick, even if the hallway between the two children is empty of people (as often happens when paranoid adults want to halt the spread of a disease). Unlike conduction and convection, the virus here isn't passed directly from person to person. Instead, it is converted into a different medium (the air) by the sneeze, and is carried by that medium until it can be converted back. With atoms, something similar happens. The energy of the atom is converted into an alternate form – in this case, oscillating excitations of the electromagnetic field, a.k.a. light – and fired off into the void. The light energy, or **radiation**, won't be converted back into heat until it hits a new atom. At that point, some or all of the energy will probably be converted into kinetic energy, which will contribute to the overall heat of whatever object the atom is part of. A great example of this is sunlight. If you've ever stood outside on a sunny day, you can feel the "warmth" of the Sun. What you're actually feeling is heat energy that was emitted by particles within the Sun and traveled across the vast and empty expanse of space. That energy remained as radiation until it struck you, at which point your atoms converted it back into kinetic energy, which you feel as heat. Neat, huh?

What's nice about radiation is that it doesn't require anything to carry it other than the electromagnetic field (which exists pretty much everywhere). That means that the Sun, for example, can heat us efficiently even though it's incredibly far away and makes no contact with us at all.

92 I don't know what it is about kids and projectile sneezing, but somehow they can fling nose boogers distances that would put most long jumpers to shame. I've never seen an adult who's able to do that, although I'm not sure if that has more to do with ability or the tendency of adults to cover their noses when they sneeze.

Here's an analogy that combines each of the three methods. Suppose you wanted to send a letter from Fairbanks, Alaska, to Miami, Florida. One way to do so would be via conduction. You hand the envelope with the letter to your neighbor to the south. He hands it off to his neighbor south of him, and so on. Each person gives the letter to someone nearby who is hopefully a little bit closer to Florida. Eventually, if we assume that no one makes any mistakes, that letter will arrive at its destination. It will take a long time, however; my estimate was approximately 7,000 hours – almost 42 weeks! This is the letter equivalent of conduction.

On the other hand, you could hand the letter to a mail carrier. That person will carry the letter with them as they move, eventually taking it all the way to the post office. There another transfer will occur, and the new person will carry the letter some distance as well. Although there are still some transfers along the way, most of the movement of the letter comes about because an individual carried it. As you can imagine, this process is much faster(less than a week, in fact, even from Alaska), and is therefore more like convection.

Finally, there's the radiation version of sending letters: email. Change the form of the message and fire it off electronically. Clearly the method is the fastest—nearly instantaneous—and when the message arrives it can be printed to reproduce the original letter. Furthermore, this method allows the letter to "travel" even across areas where no one lives. Imagine trying to use "conduction" to send a letter from New York to Paris. Without a method to send the letter across the vast emptiness of the Atlantic Ocean (either by giving it to someone to carry or by broadcasting it electronically), the letter has no way to reach its destination.

Just to prove to you that physics can be fun, I give you the following analogy that you can actually try in class (if you can convince your professor to allow it). Get six dozen poker chips – two dozen each of red, blue, and white – and divide the class up into three teams. Each team gets one color, and the team will be named based upon the color. The goal is to get the chips to the other side of the classroom as quickly as possible. I suggest a pair of bins for each team. One will contain all of the chips at the beginning of the game; the other is the bin into which the chips must be deposited in order to count.

Here are the rules of the game:

- No member of any team can have more than one poker chip in his or her possession at a time. If the team member is touching the chip, he or she is considered to be in possession of that chip.
- The teams get 30 seconds of preparation to discuss strategy and place themselves before play begins. No player may have possession of a chip until this phase is over and actual play begins.
- The white team can distribute themselves however they like before play starts. Once play starts, however, they are not allowed to move their feet. All white chips must either be in the possession of a team member, in the starting bin, or in the scoring bin.
- The red team must all begin clustered around the start bin. However, once play starts they can move as they like. All red chips must either be in the possession of a team member, in the starting bin, or in the scoring bin.
- The blue team must all begin clustered either around the start bin or the end bin. Once play starts they are not allowed to move their feet. Blue chips do not have to remain in the possession of a team member during play[93].
- Play progresses for either 30 seconds, or until one team's starting bin is empty. Each team gets three points for every chip in the scoring bin. They lose one point for every chip remaining in the starting bin.

If you imagine this game in your head, you can probably predict what will happen. The members of the white team will spread themselves out across the room. Once play begins, they will hand the chips one at a time down the line. The red team will sprint, with each person carrying one chip from the start bin to the scoring bin, and the blue team will fling the chips. Ideally the blue team will divide itself into throwers and catchers,

93 If your professor is (understandably!) nervous about the blue team, you may want to suggest playing with only the red and white teams.

but it's quite possible that the entire blue team will elect to toss chips! (This is why your professor may be nervous about letting folks actually perform this activity!)

Can you identify which team represents each of conduction, convection, and radiation?

Physical Effects of Heat

Now that we've covered how heat gets transferred, the question remains, what happens to that heat after it's been transferred? Obviously, it has to have some measurable effect. Although changes in temperature may be the more obvious of these, it's also in some ways the most complex. So I'm going to start with size changes and why they're important.

Thermal Expansion

Let's go back to that image of the mosh pit[94]. If the moshing folks are exhausted and sleepy, they won't be able to manage bouncing against each other very hard, and pressure from the outside crowd will keep the mosh pit in a small location. Now imagine each mosher is given a big cup of espresso and sugar. As they wake up and become energized, they will start moving more. More motion means they'll bounce harder, not only against each other, but also against people at the edge of the pit. These harder bounces will push back the crowd, and the mosh pit will expand. That, in a nutshell, is thermal expansion. As the material gains heat, the individual atoms gain kinetic energy. They move faster, and thus bounce against each other harder. This forces the atoms further apart, and as they spread out, the entire object gets bigger.

When most textbooks talk about the thermal expansion of materials, they use examples of cracking sidewalks, buckling steel beams, or the operation of bimetallic thermostats. However, there's an implication that is much

94 If for some reason you are not familiar with the idea of a mosh pit, it's basically an area "carved out" of the audience space at a live concert. Individuals in this area throw themselves into and bounce against each other, kind of like atoms in a liquid. The analogy isn't perfect, but it's good enough.

more vital to our existence. Consider a pot of water sitting on a hot burner. In this case, the "mosh pit" is at the bottom of the pot. The water molecules there are getting a big helping of kinetic energy. Note that I'm not interested here in how the heat spreads (although it certainly does); I'm more concerned with the water receiving that heat. As the water molecules move faster, they force themselves further apart. They bounce off of each other and the cold molecules above them, just like our coffee-chugging body-slamming moshers. Just as the mosh pit expands when the moshers gain energy, the region of hot water gets bigger as the water molecules gain energy.

However, this expansion comes with a cost! Because the same number of molecules are taking up a larger volume, the water at the bottom of the pot has become less dense! The cold water above is now heavier (by volume). The hot water has less force (weight) pulling it to the bottom of the pot than the cold water-. In this contest, the cold water wins and sinks, driving the hot water upward. The cold water enters the "mosh pit," gains some heat, expands, and moves upward as the cold water above it sinks into its place. As the now-hot water rises, it carries the heat from the bottom of the pot with it, which means that we've just described one of the most basic convection systems that exists! This convection system – a dense cold fluid sinking and displacing warmer less-dense fluid that's closer to the heat source – drives oceanic currents, atmospheric weather patterns, the motion of the tectonic plates, and the rippling granulation that keeps the Sun's surface an even temperature[95]. Not bad for the random full-body slams of a bunch of atoms, right?

It is quite possible to calculate how much a given material will expand based upon how much heat it gains. Such calculations, however, are based upon experimental results[96], and thus at this level are little more than thinly disguised algebra problems.

95 It also mixes your tea and coffee for consistency and smoothness!

96 As it turns out, it's **very** difficult to predict based upon chemical properties how much a substance will expand as it gains heat; an excellent example of this is water, which between 0°C and 4°C actually contracts as it gets hotter!

Heat and Temperature

You many have noticed that our discussion of heat has not included the "T" word – temperature. I have intentionally avoided this word because, like energy, everyone thinks they know what the word means. Unfortunately, that's not always correct. However, instead of just jumping into my definition, let's see how much you know!

*** * * POP QUIZ[97] * * ***

(choose the best answer)

1. If I add heat, then the temperature must increase.
   (a) Yes.
   (b) No.
   (c) Not if there is a phase change, but otherwise yes.
   (d) Only on days whose names end in "y" that occur during months that contain a full moon.

2. Heat and temperature measure the same thing.
   (a) True.
   (b) False.
   (c) Um, maybe?
   (d) None of the above.

3. Object A has a low temperature; object B has a high temperature. Which object has more heat?
   (a) Object A.
   (b) Object B.
   (c) The objects have the same amount of heat.
   (d) The question cannot be answered without more information.

4. How do you like talking about temperature so far?
   (a) It's hawt!
   (b) Dude, it's seriously cool!
   (c) I'm feeling kind of lukewarm about it.
   (d) Seriously? This is a question on your quiz?

97 What? I'm a certified teacher! I'm required to give pop quizzes! See page 212 of the *First Year Teacher's Survival Guide* for more information.

Time's up! How do you think you did? The answers are

1. (b)
2. (b)
3. (d)
4. (Don't judge me! The Rules[98] require at least four questions per quiz!)

Let's talk about the answers. If we add heat to an object, the temperature of the object will, in general, increase. However, there are exceptions, one of which is if the object undergoes a phase change. We'll discuss that later. Another exception, however, is if I add more material to the object as I add the heat. Consider, for example, a pot of water sitting at room temperature. If I add more room-temperature water, I'm adding heat – the molecules of the new water definitely come with some kinetic energy. However, the temperature of the water (both old and new) won't change.

These first two questions bring us to the definition of temperature. Heat is (basically) the total kinetic energy of the molecules of an object. Temperature, by contrast, is the average kinetic energy of the molecules. If heat measures how much kinetic energy all of the object's molecules are packing, temperature measures how much kinetic energy a randomly caught single molecule is likely to have. There's no guarantee that the molecule will have that much energy, of course. As the molecules run into each other, they exchange energy and therefore speed up or slow down. Because of this, there is a huge spread of actual kinetic energies possible for the molecules. However, as long as the object *overall* isn't losing or gaining heat (or molecules), the total amount of kinetic energy (heat) will remain fixed, and so will the average kinetic energy (temperature).

This is why I can increase the heat in a pot of water without changing the water's temperature. If I add both heat and molecules, the average amount of energy per molecule may remain the same. It's even possible to create a situation where adding heat could end up making the water colder – for example, imagine adding refrigerated water to that pot of room-temperature water.

98 Darn you, Fein and Schneider! Wait... Whoops, wrong rules! My bad!

That may seem somewhat counter-intuitive. After all, shouldn't heat make things hotter? This confusion is unfortunately an artifact of the English language. Hot and cold are relative terms; remember that they always imply a comparison with something else. Boiling water is cold compared to liquid aluminum. Objects in your freezer are hot compared to the surface of Saturn's moon, Titan. However, all of these things – boiling water, liquid aluminum, the stuff in your freezer, the surface of Titan – have heat. In fact, the only way for an object to have no heat is for the molecules of that object to have no kinetic energy. In other words, the only way to get rid of all of an object's heat is to completely stop all traces of motion for every single molecule within the object[99].

This actually has implications for temperature, as well. Once an object has zero heat, by definition it must have zero temperature – zero total kinetic energy means zero average kinetic energy as well. Furthermore, it's not possible to reduce the temperature further, as the kinetic energy is at a minimum[100]. After all, how do you make something move less than "not at all"? Thus, there's a minimum possible temperature, or an **absolute zero**. Actually achieving this state is more than a notion, however. As you can imagine, it's horrifically difficult to remove absolutely every single bit of energy from an object, even before you start dealing with the quantum mechanical effects that start kicking in at such extremes.

Well, if we know the bottommost temperature, then we should be able to measure upwards as we wish, right? Here's where a quirk of history messes things up. Strictly speaking, temperature is a measure of energy – average energy, but energy, nonetheless. Therefore, it should have units of energy, just like heat. However, in practice, things aren't so simple. People have been measuring temperature much longer than the microscopic view of matter has been around. Because of that temperature units seem to have absolutely nothing to do with energy. Physicists have a saying: "if it ain't broke, don't fix it," so we use the historical scales.

---

99 More thorough study of quantum mechanics further implies that it's not even physically possible to reduce a particle's motion to nothing.

100 More thorough study of statistical mechanics, quantum mechanics, and thermodynamics actually allows for negative temperatures! It depends upon a state/entropy based definition of temperature instead of the simpler kinetic energy based definition. TL;DR version: Don't worry about it right now.

The most familiar scale to readers in the United States is the **Fahrenheit** scale. Based partially upon human body temperature, the freezing point of water lies at 32 °F and the boiling point at 212 °F. These two numbers fix the entirety of the Fahrenheit temperature scale and places absolute zero at about -459 °F.

The most familiar scale to anyone outside of the United States is the **Celsius** scale. This system is based upon the freezing and boiling points of water, setting them at 0 °C and 100 °C, respectively. Based upon these numbers, the zero of the Celsius scale lies at -273 °C. A closely related (and more important) scale is the **Kelvin** scale. (This scale, by the way, is considered the international standard scale of temperature of the scientific community.) The Kelvin scale is almost identical to the Celsius scale; one Kelvin degree is the same size as one Celsius degree. The Kelvin scale, however, is based upon an absolute zero temperature of 0 K[101]. This has several nice consequences. First, all temperatures in Kelvin are positive. Second, and more important, doubling the temperature in Kelvin directly reflects a doubling of the energy in the system, and vice versa. In most other temperature scales, this is not the case. I *highly* suggest that you convert all of your temperatures to Kelvin before doing any calculations. The freezing point of water lies at 273 K, and the boiling point at 373 K.

There's one last scale, but almost no one finds this scale familiar. If you start with absolute zero as the start of your scale (like the Kelvin scale), but make one degree the same size as a Fahrenheit degree, you end up with the Rankin scale. Absolute zero in this scale is 0 °R; the freezing point lies at 491 °R, and the boiling point at 671 °R. No one uses this scale. Seriously. You shouldn't either.

As I said above, all of your calculations should be done using temperatures in Kelvin; to do this, you'll almost definitely need to know how to convert between temperature scales. Converting between Celsius and Kelvin is easy: Simply add or subtract 273. If you remember that Celsius temperatures are always smaller (numerically) than Kelvin temperatures, you should be able to figure out whether you need to add or subtract, depending on which way the conversion is going.

---

101 Unlike Celsius or Fahrenheit, the temperature in Kelvin is not marked with the degrees (°) symbol.

Just in case you don't see how that works, here's an example. Let's say you're given a temperature of 45 °C that you need to convert to Kelvin. The temperature in Kelvin should be bigger, so add 273 and you're done.

$$45\,^{\circ}\mathrm{C} = 318\mathrm{K}$$

Similarly, a temperature of 512K should be a smaller number when converted to Celsius, so subtract 273:

$$512\mathrm{K} = 239\,^{\circ}\mathrm{C}.$$

Easy!

The conversion between Celsius and Fahrenheit is a little bit more difficult. I won't throw that annoying 5/9ths equation (or is it 9/5ths?) at you; I can never remember which way the fraction goes or whether I should add (or subtract!) the 32 before or after I multiply... Ugh! There must be a simpler way!

Fortunately, there is, and the only thing you need to remember is that the temperature measured in Celsius tends to be smaller (numerically) than the same measurement in Fahrenheit.

i. Take the number you want to convert, and add 40.
ii. If you're converting Fahrenheit to Celsius, divide the number by 1.8. (Division will make the number smaller, which is what you want for a result in degrees Celsius.) If you're converting Celsius to Fahrenheit, multiply by 1.8. (Multiplication will make the number bigger, which is what you want for a result in degrees Fahrenheit.)
iii. Subtract out the 40 you added in originally, and you're done!

Strictly speaking, the above steps replicate that annoying pair of conversion equations. For example, 1.8 is the same as 9/5ths or 5/9ths, depending on whether you're multiplying or dividing. However, I find the above steps easier to remember and execute from memory than trying to remember the traditional methods. I don't know who originally came up with this method, but I give that person my hearty and sincere thanks.

So, as an example, let's convert 68 °F to Celsius.

i. Adding 40 gives 108.
ii. Dividing by 1.80 gives 60. (Remember, the number in Celsius should be smaller than the number in Fahrenheit.)
iii. Subtracting the original 40 gives 20, which is our answer in Celsius: 68 °F = 20 °C.

Converting from Fahrenheit to Kelvin (or vice versa) will require a layover in Celsius.

Specific Heat, Latent Heat, and Loss of Atmosphere

Whew! Temperature is far more complex than most people realize, and we haven't even scratched the surface. As I said, the temperature of an object measures the average kinetic energy of the molecules that make up that object. So, adding heat (and only heat) to a substance should increase the temperature, right? By how much?

As you might expect, the answer to this question depends on more than just how much heat you add. For example, a molecule of water consists of two hydrogen atoms tied to an oxygen atom. Heating a bucket of water will make each of the individual water molecules move as a unit; however, it will also make the hydrogen atoms move relative to the oxygen atom. The hydrogen atoms will have trouble escaping, however, so this motion will simply cause them to shake and jitter around the oxygen like tiny bobbleheads. While this energy is part of the overall heat in the bucket of water, it does not contribute to the average kinetic energy of the water molecules. In other words, although the amount of heat in the water increased, only part of that heat contributed to increasing the temperature. Some of it got stuck making the parts of the molecule itself vibrate[102]. The amount of heat necessary to increase an object's temperature by a given amount depends on what the object is made of!

102 These alternate places that energy can go instead of the overall linear motion of the molecule are called modes. Generally speaking, the more ways the pieces of a molecule can move relative to each other the more modes the molecule has, and the more heat it will take to increase the temperature of that substance by one degree. Modes are sometimes called **degrees of freedom**.

Unfortunately, there's a second complication. Not only does heat get trapped in how the atoms of the individual molecules move, it can also get stuck between the molecules. Remember the analogy with the mosh pit? Imagine each of the moshers has strong BO[103], and that every mosher hates the odor of every other mosher. As they approach each other, the smell gets stronger and stronger. You might imagine that the moshers end up recoiling from the smell well before they make actual contact and bounce off of each other. The smell, then, creates a second, weaker "inter-mosher" interaction. It's not enough to completely negate the collisions, but it definitely affects how the moshers move. Molecules experience similar kinds of interactions[104]; however, these are much more complex. Like atomic vibrations these intermolecular interactions can take a share of the heat that goes into the object.

In addition, these interactions can constrain how much the molecules can reasonably move. Think of a solid, for example. The (relative) lack of heat constrains how the molecules can move; this contributes to the object's rigidity and structure. As heat is added the motion of the molecules increases, and the temperature of the object increases. Eventually, however, the intermolecular interactions limit how much the molecules can move. If the molecules want to move faster, the nature of the interaction has to change. Additional heat goes into changing the interaction, rather than speeding up the molecules. This is called a **phase change**; the solid becomes a liquid. Curiously, this process effectively occurs all at once[105]; the temperature of the one phase cannot change until the other phase is completely gone.

What does all of this mean? Essentially, that while in principle it's possible to predict how the heat will be divvied up based upon the chemical properties of different elements and how they interact, in practice it is painfully difficult. Instead, scientists work with **specific heat** and **latent heat**. The specific heat of a material is how much heat energy it takes to raise

103 Hopefully this is not a big stretch of your imagination. By the way, if you don't know the initials, "BO" stands for "body odor".

104 For example, van der Waals interactions.

105 For our purposes, that is; if the object is big enough or the heat is added fast enough, strange things can happen. You generally don't run into that sort of thing in introductory physics, however.

the temperature of some specific amount of that material by one degree[106]. These values are usually determined experimentally and added to the big table o' specific heats. You can use the big table o' specific heats to actually calculate out how much of a temperature change occurs in an object from the amount of excess heat added to the object, if you know how much of the object you have and what the object is made of. The total heat energy contained in such an object (Q) at a given temperature (T) can be calculated using the specific heat for the material (c) and the object's mass (m):

$$Q_{current} = m\,c\,T_{current}$$

Latent heat, on the other hand, deals with those points of phase change; the latent heat of a material is how much heat energy it takes to change a specific amount of that material from one phase to a different phase at the same temperature[107]. Again, these values are determined experimentally; there's a different table for them, however – the big table o' latent heats. The amount of heat it takes to completely change the phase of a given amount of material can be calculated using the appropriate latent heat ("L"):

$$Q_{phase\,change} = m\,L_{by\ material,\ by\ type\ of\ phase\ change}$$

The latent heat of fusion[108] (or freezing) for water is different from the latent heat of vaporization (or boiling) for water; the latent heat of vaporization for carbon dioxide is different to the latent heat of vaporization for water. Make sure what you have both the right material and the right change of phase when you're calculating this heat.

---

106 In physics classes the specific heat c almost always comes in SI units – joules per kilogram per Kelvin degree, or J/kg·K

107 In SI units latent heat "L" comes in joules per kilogram, or J/kg. There's no reference to temperature because the temperature can't change until the entire object has changed phase.

108 You may also see references to the "latent heat of condensation" or even "latent heat of melting". These values should be the same as the latent heat of vaporization and the latent heat of fusion, respectively; they're referring to the same phase change in the opposite direction. In other words, condensation is the "undoing" of boiling (vaporization), and melting is the "undoing" of freezing (fusion).

As you can see, this means that figuring out temperature changes for a given material can be quite complex. A given chemical compound usually has three specific heat values (one each for the solid, liquid, and gaseous phases) and two latent heat values (one for the change from solid to liquid, and another for the change from liquid to gas). In addition, you have to know the temperature at which each of the phase changes occurs! Again, these values are almost always determined experimentally. That's seven different numbers to describe one material! Imagine replicating that for every single possible material, and you can see how complicated this can quickly become, before we even consider mixing stuff together Thankfully, in the end we're only concerned about heat, which means although it seems like we've gone around the world and back, we're still only talking about energy. The same scaffold that will help us solve problems of objects falling onto springs can easily help us solve problems of temperature and phase changes. We only need one more definition to do so. **Thermal equilibrium** is, in simple terms, when two objects are at the same temperature. If two objects are allowed to interact – i.e., they're in contact – then they will eventually come into thermal equilibrium. The heat energy will spread out evenly through all of the accessible material, and they will end up at the same temperature. With that final point we can solve all kinds of problems involving temperature and heat. I think we're ready to review the energy scaffolding now. In the next chapter– what? Loss of atmosphere? Oh, right! I mentioned that in the heading to this section. I really should say something about that, shouldn't I?

As I said before, temperature is a measure of the average kinetic energy of the molecules of an object. The kinetic energy, in turn, is related to both the mass and the speed. As the temperature of a gas[109] increases the average kinetic energy of its molecules increases; as the kinetic energy of a molecule increases its speed increases. The hotter the gas, the faster (on average) its molecules move.

Now consider a mixture of two different gases. If you wait long enough, the two gases will be in thermal equilibrium; they'll have the same temperature. But different gases are made of different compounds; the mo-

109 Although it may seem strange, a bunch of gas can be considered an object! In fact, for our purposes any big collection of gas (like the air in a balloon, our atmosphere, or the Sun) will be considered an object.

lecules of those compounds will almost always have different masses. At the same temperature all of the molecules will have the same average kinetic energy; however, for the heavier (or more correctly, the more massive) molecules the speed will be lower. The higher the molecular mass, the lower the average speed. At the same temperature, the molecules of the less massive gas will move faster than those of the more massive gas.

As it turns out, this has huge implications. Planets (including the gases that surround them) are held together by gravity. An object moving away from a planet experiences a gravitational force that pulls it back towards the center of the planet. However, this force decreases as the distance gets bigger. If an object is moving fast enough, it won't stop – it will just continue out and away, and not return[110]. There's a cutoff speed, called the **escape velocity**, that is determined by the size and mass of the planet. Any object that is moving slower than the escape velocity will eventually turn around and return to the planet; any object moving faster won't. It may slow down, but it will never stop.

What does this mean for gases in an atmosphere? First, hot planets are much less likely to have atmospheres than cold planets. Remember, the hotter the gas, the faster the molecules move. A hot atmosphere has fast moving molecules; the planet with that atmosphere will need to have a very high escape velocity to keep those molecules around. This is one of the reasons that astronomers didn't expect to find the so-called "Hot Jupiters" – gas giants orbiting closer to their stars than Mercury orbits the Sun. Lots of light from the star means that the atmosphere will be very hot; that increases the likelihood that the atmosphere will escape.

Second, gases with low molecular mass will escape before gases with higher molecular mass. On Earth, for example, hydrogen and helium move much faster than Earth's escape velocity, so our atmosphere does not contain much. Jupiter, on the other hand, is further from the Sun (and thus colder); it's also has much more mass, and thus a larger effect on the gravitational field. Jupiter's atmosphere is predominantly hydrogen and helium. By contrast, both oxygen and nitrogen have comparatively high molecular masses; Earth has no trouble hanging onto these gases. It's no co-

110 In other words, what goes up *might not* come down, depending on how fast it's going up.

incidence that over 99% of molecules in the atmosphere contain at least one of these two elements[111].

One interesting consequence of this is the potential loss of surface water. As the temperature of the planet increases, more water evaporates. As the water vapor drifts higher and higher it can interact more and more with energetic particles and light from the Sun. These high energy interactions can break the molecules of water into elemental oxygen and hydrogen. Guess what happens to the hydrogen at that point?

Before you panic, please realize that the amount of water lost in this manner as a fraction of the total amount of water on Earth is negligible. However as the Earth's surface temperature increases (for a variety of reasons) it is likely to become a more dominant method of loss. Many planetary astronomers believe that Venus once had water reserves much like Earth does now. Over time, however, the surface temperature increased, and as more and more water evaporated more and more was lost to similar high energy interactions at the top of its atmosphere. In other words, an understanding of the interactions between temperature and molecular mass can actually help explain why the planets look the way they do!

Man, that's cool. This is why I love science. You think you're going to learn about one thing, and you end up learning about a bunch of other related stuff as well. Awesome!

Okay! I think we're ready to see the energy conceptual scaffold now. Let's solve some problems!

111 You may know that there's about four times as much nitrogen in the atmosphere than oxygen; partially this is due to how much more chemically reactive oxygen is compared to nitrogen. In other words, oxygen loves hanging out and bonding with other elements; nitrogen, not as much.

Energy: Conceptual Scaffold

As it turns out, problems based upon energy are generally much easier than ones based upon forces. As such, the conceptual scaffold for energy is simpler as well:

1. Prepare
2. Determine the total initial energy
3. Determine the total final energy
4. Apply conservation of energy
5. Do math and solve

The rest of this chapter explains each step in more detail. In the next chapter, I'll show you the conceptual scaffold in action by applying it to several different example problems.

First, here's an abbreviated remix of the reminder from the forces conceptual scaffold chapter. First, the conceptual scaffold isn't a solution, just like construction scaffolding isn't a building. Although the scaffolding will give you a set of steps to follow, you'll still need to bring your knowledge and understanding to bear to make decisions for whatever specific problem you're trying to solve. Your knowledge of the physics will (hopefully) help to inform and guide your decision-making as you use the scaffolding. Second, every step of the conceptual scaffold is designed to use what came before to prepare for what lies ahead. To successfully complete a given step, you will need to have completed the previous steps correctly; successfully completing a given step means that you're prepared to successfully complete the steps that lie ahead. To return to the construction analogy, you can't build the third floor before you build the first and second, but once you've built the first and second floors, you're ready to tackle the third. Finally, the last step, "Do math and solve," is simultaneously the most important and least important step. That step is the one where you finally answer the question that was posed.

1. Prepare

Just because energy problems are often easier than force problems doesn't mean you shouldn't take a moment to ready yourself for the work ahead. Make sure you have a pencil, an eraser, and maybe a calculator. Perhaps a drink, a small bag of chips, or some fruit? Possibly a sweater, in case you get cold, or a pillow if your seat becomes uncomfortable? A second pair of socks? Oh, a spare pair of pants in case you spill your drink! And sunscreen! Definitely sunscreen; if I don't take that, I'll burn. And what about –

Wait, I'm preparing for a trip to a Caribbean island. Sorry about that. A pencil and eraser (and maybe a calculator) should be enough. Oh, and paper. You'll probably need some paper. But that's it[112].

Then, finally, read the question. Once you've read it, take a moment to imagine the situation described, and study any picture provided. Then, read the question again. If there is anything you don't understand, ask about it. Getting clarification is not the same thing as asking for the answer. If necessary, take a moment to relax again.

You're ready to rock and roll![113]

Looking Forward: Make sure you understand the situation as described by the problem statement. If you're having trouble imagining what's going on, you should get more information.

112 I'm pretty sure that's it. Generally, if you want a suitcase to do physics problems then you're probably confusing “working on physics” and “planning an escape to the tropics”. It's a common mistake.

113 **Disclaimer:** although you may be ready to rock and roll, it would be more prudent and more efficient to work on physics. Once you're done with the physics you can rock your socks off. Figuratively speaking, of course; friction doesn't really respond much to music. Please don't be disappointed if you rock out loud and your socks remain on.

2. Determine the total initial energy

Energy problems always involve a change. If you remember, the whole point about the concept of energy was the idea that it gets passed around like money. If there's no change, and the energy isn't changing forms, then it's like that cash in the closet – it's just not that interesting. So the whole point of conservation of energy problems is to figure out who had the stuff right before the change, and who had it right afterward. So to start with, you need to figure out who's involved before the change. Kinetic energies are generally the easiest to identify. Is anything moving? If so, you've got a kinetic energy! The way in which the object is moving will determine what kind of kinetic energy you should add.

Internal energy is almost as easy. If temperature or friction is mentioned, then you will probably have to deal with heat. If a spring, rubber band, or other elastic material is mentioned, then you will probably have elastic energy to worry about. Other forms of internal energy will probably be explicitly mentioned – watch for references to batteries, gasoline, fuel cells, hydrox tanks, radioisotope thermoelectric generators, miniature aneutronic nuclear reactors, antimatter power conversion units, and so on. Even food, food substitutes, and imitation food substitutes[114] could be considered sources of internal energy, depending on the problem!

Potential energies are often difficult to identify, mainly because problem statements rarely explicitly mention the field itself. Fortunately, there's a built-in limitation. Only two fields tend to show up in an introductory physics class: the gravitational field and the electric field. If the problem takes place near the Earth (or Mars, or the Moon) then you should add the gravitational potential energy. If the problem mentions positive or negative charges, then you should add the electric potential energy.

Finally, it's possible that either the problem includes a force or some other source is adding energy to the problem. If that's the case, give them a name (like "E_{source}", or whatever makes sense to you), add them in, and move on.

---

114 Not imitation foods, though. Nobody likes those guys – not even other imitation foods.

As always, the point here isn't to calculate the actual number for any of these energies. The number doesn't mean much anyway – after all, what does 876.3J mean[115]? Instead, simply write each form the energy could take – linear kinetic energy, electric potential energy, heat, whatever. This captures the essence of what's going on, and leaves a clear record of your thought process while solving the problem. Besides, our last step is "Do Math and Solve"! Don't steal his job, man!

Once you have a list of all of the different forms the energy is in before the change, add them up. That's the total initial energy.

Looking Back: Read the problem statement, and make sure you haven't missed anything. Is there anything else going on before the change that isn't represented? Make sure you caught everything!

Looking Forward: Eventually we will want to calculate the values for each type of energy you've identified. If you have this equation handy (or you just know it by heart), you can write it down so that you have it accessible for Step 4. This isn't absolutely necessary, however.

3. Determine the total final energy

Now that you know where the energy started you need to know where it finally ended up. Think about the change that occurred. What happened, and what was the result? Did an object's movement change? Did the object change location? Was fuel burned, or a battery depleted? Did the temperature increase? Each of these implies that a different particular energy should be listed.

For the most part, this step proceeds in a very similar fashion to the previous step – go through the different types of energy and figure out whether each type applies to the situation or not. Once you're done, add them up. The total your total final energy.

115 Not much if you're a planet (even a small one). A whole heck of a lot if you're a proton or electron!

Looking Back: Look over the energies you listed for Step 2. All of those are fair game for this step, even if you're fairly sure that one or all of them were drained by the change like a hemophiliac at a vampire's birthday party. Don't feel bad about adding them to the list, even if they later turn out to be zero.

Looking Forward: As in Step 2, you may want to write down the equations for each type of energy you found, if they are easily accessible. If they're not, don't worry about it. You can get them later.

4. Apply conservation of energy

Remember, conservation of energy states that energy can't mysteriously appear out of nowhere, nor can it vanish into nothing. This means that any energy that we had before the change has to be accounted for after the change as well. There may have been some shuffling of how the energy is distributed, but we should be able to track all of it. The initial total energy you found in Step 2 should equal the final total energy you found in Step 3.

Go ahead and set them equal.

Yes, this step is really that easy, although there is one warning I will give you. If you have two or more objects that interact (they run into each other, hang out, get a cup of coffee, look at each other meaningfully, etc.), then this problem isn't really a conservation of energy problem! (D'oh!) At best, it's a collision, in which case you'll need to jump forward to the section on conservation of momentum and use the conceptual scaffold there[116]. Don't fret, though; there's a good chance you'll still need what you've done so far.

You're now ready for the last step!

116 At worst, you've mistaken a romance novel for your physics test and therefore you have bigger problems to handle than which conceptual scaffold you should be using.

Looking Back: The real work of conservation of energy problems is in correctly determining the initial and final total energies. Even then, however, they're usually pretty straightforward.

Looking Forward: If you haven't written down the equations to calculate the values for each type of energy, now is the time to grab them. You'll need them to solve the problem!

5. Do math and solve

Now that you have an equation, look at what the question statement is requesting. What energy is that associated with, and where is it in the equation? Solve for that thing and plug in the numbers!

Conservation of energy problems are generally very straightforward. If this seemed deceptively easy, it's probably because it was. Because of this, energy problems are sometimes "stacked" – they have multiple parts (usually labeled alphabetically) and each part's initial or final setup is related to the final or initial setup of one of the other parts. For example, you might have a block that slides down a ramp (part a) then slides to a stop on a horizontal rough surface to generate some heat (part b). Just use the energy conceptual scaffold repeatedly. The only difference will be how you choose what counts as before the change, and what counts as after.

Looking Back: If at any time you are unsure of where a particular term in your equation came from, you should be able to track backwards through the steps to identify its source. This is the strength of being careful and systematic. Everything should be linked both forward and backward in a logical way. If something comes out differently than you expect, take the time to try to work backwards to figure out why. Once you have an answer to the question, consider whether or not it was what you expected. Does your answer make sense given the problem statement and the energies you chose?

Congratulations! You're done!

Energy: Examples

Here you will find a few sample problems using the energy conceptual scaffold. I've set the problem statements here so that you can try applying the conceptual scaffold by yourself, then check your work against the example solutions.

Example 1: A 102.0408 g ball is loaded into a spring launcher (k = 248.1632 N/m) that is pointed straight upward. The spring is then compressed 12.4680 cm before being fired. How high does the ball go relative to the mouth of the launcher?

Example 2: Santa Claus is out driving one night at 70.0 mph when a reindeer jumps out onto the road in front of him. He slams on the brakes, and the car slides to a stop, narrowly missing the reindeer. The tires on Santa's car account for 2.5 percent of the car's mass. Assuming that half of the energy released ends up in the tires, how much does the temperature of the car's tires increase in Celsius degrees? The specific heat for the car's tires is 1250 J/(kg K).

Example 3: The Margerums are world-renowned moose tamers, and Will is by far the best of the bunch. His secret recipe of moose feed is known to be perfectly irresistible, as long as it's mixed properly and delivered at the perfect temperature of 15.0 °F. He keeps the dehydrated feed stored in a temperature-regulated deep freezer set at -12.0 °F, then reconstitutes it carefully and slowly, using water at exactly 67.2 °F. To feed the four calves in his stable (Mickey, Mini, Jerry, and Danger),Will needs to prepare 24.0 kg of dry feed. How many liters of water does he need to bring the feed to the proper temperature for the calves? Assume that the specific heat of the dry feed is 1820 J/(kg°C), and that all of the water will be absorbed by the feed during the warming process. Also, what is the final weight (in pounds) of the reconstituted moose feed?

<u>Example 1</u>

A 102.0408 g ball is loaded into a spring launcher (k = 248.1632 N/m) that is pointed straight upward. The spring is then compressed 12.4680 cm before being fired. How high does the ball go relative to the mouth of the launcher?

1. Prepare

Just as with forces problems, you'll need something to write with and something to write on. Pen and paper? Computer program? Even wood ash and a cave wall will work, although in that case I suggest you be prepared to spread out some.

We must also take a moment to envision the situation described in the question. Although energy problems don't require a picture, in some cases, like this one, it can help clarify where things are. For example, there are three locations in this problem that could be important: the location of the ball when the spring is compressed, the location of the ball when it's at its highest point, and the mouth of the launcher(this one is explicitly mentioned in the problem!). So a drawing here can help clarify where these locations are, relative to each other.

I've included such a picture here. Note that there are several distances we may need to consider: h, h_c, and x. In the next couple of steps, we'll look at why each of those is necessary.

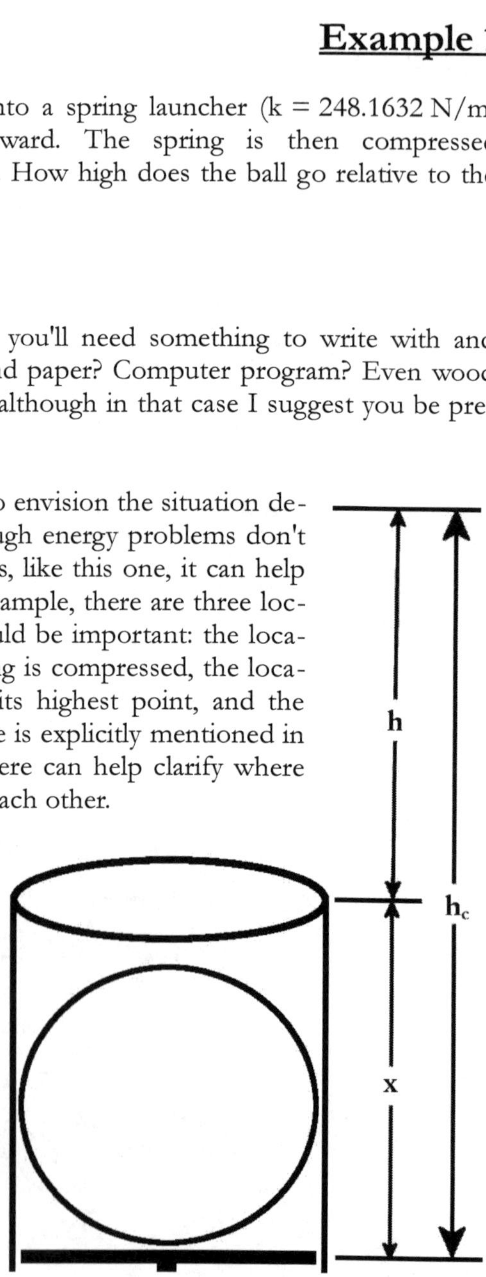

2. Determine the total initial energy

Time to figure out where the energy is initially. To do this, let's consider what is explicitly mentioned in the problem. We know there's a ball, and presumably the ball will be moving, so we might have kinetic energy. The problem also mentions a spring launcher, so maybe we'll need to deal with some elastic internal energy. The only kind of energy that's not explicitly mentioned is potential energy. However, remember what I said when I introduced the energy conceptual scaffold:

"Potential energies are often difficult to identify, mainly because problem statements rarely explicitly mention the field itself. ...However, there are only two fields that tend to show up in introductory physics: the gravitational field and the electric field. If the problem takes place near the Earth (or Mars, or the Moon) then you should add the gravitational potential energy. If the problem mentions positive or negative charges, then you should add the electric potential energy."

Well, the problem didn't mention any charges, so we can safely rule out the electric field. What about the gravitational field? When I imagine a spring-loaded launcher, I imagine a foam dart gun or some other toy, not some exotic extraterrestrial deep space thing. In other words, I imagine the launcher existing here on Earth, and nothing in the problem statement leads me to believe that this assumption is incorrect[117]. So I should probably include the gravitational potential energy as well.

Still, we must ask ourselves, do all of these actually come into play? The ball starts in a compressed spring launcher, as shown on the previous page. At that point, the ball is sitting there, waiting to be fired. That means it's not moving! (Yet.) So the ball's initial kinetic energy is zero. The spring, however, has been compressed – the problem statement specifically states that! Even more, it's been compressed by 12.4680 cm. That's the distance x in the picture. So we know there's elastic internal energy available.

The gravitational potential energy depends on the height from the ground. Unfortunately, we just don't know where the ground is! As it

117 If you did imagine some exotic extraterrestrial deep-space thing, let me know! I love hearing from crazy peop– err, I mean, alternative reality thinkers.

turns out, this isn't as much of a problem as it might seem. The "ground" for measuring height is simply the closest low point that's convenient. If you'd like an example of this, consider measuring your height. Your height doesn't change when you measure it from the floor, regardless of which floor you're measuring from (first, second, third, etc.) All that matters is that you measure from the floor your feet are touching! In the same way, we can choose any point to measure our height from, as long as we measure from that point for all heights.

Here, we have several points we could choose to measure from: the lowest point, where the ball is compressed; the highest point, where the ball stops moving; or the middle point, at the mouth of the spring launcher. This is mostly a matter of taste, although – like most things in physics – certain choices can make the problem easier or harder. In most cases, you can simply use the lowest point as zero, and you'll be . Therefore, our height will be measured from the ball's location when the spring is compressed. This means, then, that our initial gravitational potential energy is zero, because the ball (by choice[118]) starts at zero height. It appears that the only initial energy we need to consider is the elastic internal energy:

$$E_{initial} = \frac{1}{2} k x^2$$

One thing to note, however, is that x, the spring compression distance, is not zero when the height is zero!

3. Determine the total final energy

We're interested in how high the ball will go, so our final point is when the ball has reached the top of its arc. How do we know when that occurs? It's no longer moving upward, but hasn't started moving downward yet – in other words, it's the split second in the air when the ball isn't moving[119]. If the ball isn't moving, it has no kinetic energy. In addition, the spring is

118 If this makes you uncomfortable, remember that the only thing that matters is how the energy changes from the initial point to the final point.

119 However, it is still accelerating! It was moving before that moment and it will be moving again after that moment. But at that moment, it's not moving. Hopefully that thought didn't make your head explode, but if it did.... BOOM (conceptual) headshot!

no longer compressed, so there's no energy there anymore. The only possibility left is the gravitational potential energy. This actually makes sense; if the ball is at the top of its arc, it should be higher than it was before it was launched.

In other words, the internal energy of the spring was converted to the kinetic energy of the ball. As the ball rose, however, that kinetic energy was converted into gravitational potential energy. At the top of the arc, all of the energy is in the gravitational field. At that point, the ball has gone up a height h_c, based upon the point we chose as zero in the last step and the picture we drew back in Step 1.

$$E_{final} = mgh_c$$

If we look forward just a little bit, we'll notice that the height requested in the problem statement ("... relative to the mouth of the launcher") is *not* the height we have here. We'll need to account for that in Step 5.

4. Apply conservation of energy

Conservation of energy states that there can be no gain or loss of energy; whatever we started with has to equal what we ended with:

$$E_{initial} = E_{final}$$

Thus, we can set the total *initial* energy we found in Step 2 equal to the total *final* energy we found in Step 3:

$$\tfrac{1}{2} kx^2 = mgh_c$$

That's pretty much all there is to it.

5. Do Math and Solve

At this point, all we have left is algebra – solve for the unknown:

$$h_c = \frac{k}{2mg} x^2$$

You're thinking we're done, right? Not quite! Back in Step 3, I pointed out that the problem asked for the final height of the ball, relative to the mouth of the launcher, *not* the point at which the spring was fully compressed. At this point, we have two choices: We could redo the problem completely, choosing this time to put the zero height at the mouth of the launcher, or we could simply figure out the correct height from the solution we already have.

If you choose to do option one, go ahead and work it out. I'll wait. (Remember that any gravitational potential energy that's lower than your zero point will be negative!)

Okay, done? You can check your answer against option two, which is what I will do here. Looking back at the picture from Step 1, it's obvious that the only difference between the compressed location and the mouth of the launcher is the distance x. If we subtract that from the height we found, we'll have the answer!

In equation form,

$$h = h_c - x = \frac{k}{2mg} x^2 - x$$

Converting all the numbers to base units and plugging them in gives us

$$h = 1.80222 \text{ m}.$$

Booyah! Another physics problem satisfyingly solved!

Example 2

Santa Claus is out driving one night at 70.0 mph when a reindeer jumps out onto the road in front of him. He slams on the brakes, and the car slides to a stop, narrowly missing the reindeer. The tires on Santa's car account for 2.5 percent of the car's mass. Assuming that half of the energy released ends up in the tires, how much does the temperature of the car's tires increase in Celsius degrees? The specific heat for the car's tires is 1250 J/(kg K).

1. Prepare

From here on out, I am going to assume that you know to get stuff to record your work[120] and dive right into the real prep work. First, make sure that you understand the situation that the problem describes. There's Santa behind the wheel of his convertible, cruising along and enjoying his vacation time, when suddenly Rudolf leaps out of the trees, running from the other reindeer who are all too happy to play "Reindeer Games"[121] with him now. The rest, as they say, is the problem statement. Do the tires on Santa's summer ride melt like the wax of a lit Christmas candle? Or is Santa the only thing steaming after his jarring stop?

Unfortunately, while the mental picture is entertaining, it's not particularly helpful. We don't know how far Santa stops from the reindeer (other than the miss being "narrow"), so it's just not possible to draw a picture. If you decide to draw one anyway, please send it along. I'd love to see your creative imaginings of the situation.

2. Determine the total initial energy

Although we're missing some information, it's still pretty clear that we have some kinetic energy; after all, Santa and his car are moving at highway speeds. We don't have to worry about potential energy. Although we can assume that this all takes place on Earth, the problem statement

120 This is totally not related to me running out of jokes about preparation. In reflection, however, maybe I should have prepared more...

121 Of course he's seen the movie, which is why he's running for his life.

doesn't mention the car moving up a hill or down into a valley. Why is that important? Well, remember that gravitational potential energy only changes when the object's height changes. If the car doesn't change height – say, by going up a hill or down into a valley – then its gravitational potential energy won't change either. If we know that it won't change from the initial point to the final point, we don't have to include it[122]. Similarly, there are no electric fields or charges mentioned, so we don't have electric potential energy either.

That said, the only other possible type for the initial energy is internal energy. The problem is asking about a change in temperature of the tires, so we almost certainly need to consider the tires' heat energy. Other than that, we aren't really given enough information to consider including any other kinds of energy. Let's go with what we have – the car's kinetic energy and the heat in the tires – and see how it goes:

$$E_{initial} = \tfrac{1}{2} m_{car} v_{car}^2 + m_{tires} c_{tires} T_{i,\,tires}$$

If we've missed something, we'll most certainly figure that out in one of the later steps.

3. Determine the total final energy

Santa's car comes screeching to a stop, so at that point he has no more kinetic energy. The problem statement seems to assume that it will all be converted into heat, and it explicitly states that half of that heat will end up in the car's tires. That means we've effectively been given our final total energy!

$$E_{final} = m_{tires} c_{tires} T_{f,\,tires} + Q_{other}$$

Now, you may be wondering, "where did the other half go?!" or "what's Q_{other}?" It doesn't actually matter. All we need to know is that the tires get half of the car's kinetic energy. That means that wherever the other half ends up, the amount is equal to half of the initial kinetic energy.

---

122 You CAN include it, if you like, but it will simply cancel out when you do the math.

In other words,

$$Q_{other} = \tfrac{1}{4} m_{car} v_{car}^2$$

To keep everything clear, however, we'll wait until the last step to plug this in.

4. Apply conservation of energy

As always, we can't create or destroy energy; what we started with must be the same as what we ended with:

$$E_{initial} = E_{final}$$

Therefore, we can set the total initial energy we found in Step 2 equal to the total final energy we found in Step 3:

$$\tfrac{1}{2} m_{car} v_{car}^2 + m_{tires} c_{tires} T_{i,\,tires} = m_{tires} c_{tires} T_{f,\,tires} + Q_{other}$$

Again, it really is this quick and easy!

5. Do Math and Solve

At this point, all we have left is algebra. First, let's use the insight we gained in Step 3 and eliminate that annoying "Q_{other}":

$$\tfrac{1}{2} m_{car} v_{car}^2 + m_{tires} c_{tires} T_{i,\,tires} = m_{tires} c_{tires} T_{f,\,tires} + \tfrac{1}{4} m_{car} v_{car}^2$$

Awesome! Now our equation only references things we either have or want. Here we can use another trick. The problem statement asks only for the increase in temperature – in other words, how much the temperature changes. $T_{final} - T_{initial}$ is the change in temperature that we want, so go ahead and solve for that difference:

$$T_{f,\,tires} - T_{i,\,tires} = \frac{m_{car} v_{car}^2}{4\, m_{tires} c_{tires}}$$

We don't know the mass of the car or the tires, but we don't need to. The problem statement gives us the ratio of how they relate:

$$\frac{m_{tires}}{m_{car}} = 2.5\% = 0.025$$

We can use this to eliminate both the mass of the car and the mass of the tires – isn't that convenient? Plugging this into our equation for the temperature change and simplifying gives

$$T_{f,tires} - T_{i,tires} = \frac{10 v_{car}^2}{c_{tires}}$$

We know the car's initial speed and the specific heat for the tires, so we're finally ready to calculate! Almost. First we need to convert miles per hour (mph) into meters per second (m/s). There are a variety of ways to do this conversion, but I'm at a computer so I'm going to cheat and use Google[123]:

70.0 mph = 31.3 m/s

When we plug everything in and turn the crank, we find that the tires' temperature increases by 7.84 Celsius degrees. Considering that tire temperatures can increase by up to 30 Celsius degrees simply by driving on them[124], this doesn't seem terribly alarming.

Done!

123 Make sure that you know the basics of doing these kinds of conversions by hand, as I doubt your instructor will be okay with you looking it up online during an exam.

124 The actual temperature increase depends on quite a few things, including the age and inflation pressure of the tire.

Example 3

The Margerums are world-renowned moose[125] tamers, and Will is by far the best of the bunch. His secret recipe of moose feed is known to be perfectly irresistible, as long as it's mixed properly and delivered at the perfect temperature of 15.0 °F. He keeps the dehydrated feed in a temperature-regulated deep freezer set at -12.0 °F, then reconstitutes it carefully and slowly, using water at exactly 67.2 °F. To feed the four calves in his stable (Micky, Mini, Jerry, and Danger), Will needs to prepare 24.0 kg of dry feed. How many liters of water does he need to bring the feed to the proper temperature for the calves? Assume that the specific heat of the dry feed is 1820 J/(kg C°), and that all of the water will be absorbed by the feed during the warming process. Also, what is the final weight (in pounds) of the reconstituted moose feed?

1. Prepare

This is another problem where a picture is not likely to help. However, note that the temperatures we're given are all in Fahrenheit, whereas the specific heat of the feed is in Celsius degrees. Because we'll need to convert a bunch of temperatures, it's a great time to demonstrate the temperature conversion trick I described in the chapter on heat:

Given	Add 40	Divide by 1.8	Subtract 40
15	55	30.56	-9.44
-12	28	15.56	-24.44
67.2	107.2	59.56	19.56

The numbers in the first column are those given in Fahrenheit; the numbers in the last column have been converted to Celsius.

Please note that while it's necessary to convert these temperatures, we could have just as easily left this for Step 5.

125 A moose once bit my sister...

2. Determine the total initial energy

This problem is a bit different from the previous two in that the only energies involved are internal energies – heat. The problem describes a situation where the heat stored in warm water is used to increase the temperature of cold feed so that the equilibrium temperature of the mixture is somewhere in between. So we'll start with heat that's in the frozen feed and the warm water:

$$E_{initial} = m_{water} c_{water} T_{i,\,water} + m_{feed} c_{feed} T_{i,\,feed}$$

No motion, no changes in height, no charges or electric fields, and no springs – there's no other pertinent energy that will come to play. Onward!

3. Determine the total final energy

Once the mixture has come to thermal equilibrium, everything should be at the same temperature:

$$E_{final} = m_{water} c_{water} T_f + m_{feed} c_{feed} T_f$$

Again, the only energies we need to worry about are the heat contents of the mixed water and feed.

4. Apply conservation of energy

By now, you know the drill: What we start with has to be the same as what we end up with. We set our initial total energy equal to the final total energy to get:

$$m_{water} c_{water} T_{i,water} + m_{feed} c_{feed} T_{i,feed} = m_{water} c_{water} T_f + m_{feed} c_{feed} T_f$$

I will point out that although this equation looks kind of complicated, it should be easy to look back and see where each term came from. Ensuring that the math says the same thing as the physics is the primary reason to consistently use the conceptual scaffolds.

5. Do Math and Solve

As always when we reach this step, all we have left is algebra. The problem statement asks for the amount of water necessary to make the feed, so let's solve for the water's mass. By the way, you may notice that this is the only thing we don't know – if you remember to look up the specific heat of water[126].

However, I'd like to pause at an intermediate step for a moment. What does the following equation say?

$$m_{water} c_{water} [T_{i,water} - T_f] = m_{feed} c_{feed} [T_f - T_{i,feed}]$$

Consider this carefully before checking out the answer[127]! You can pass introductory physics without learning to read mathematical equations. However, you'd be missing out on a valuable (and quite marketable) skill. In addition, if you plan on doing any sort of work in the sciences or engineering, you'll find that learning to interpret equations will become a necessity.

Anyway, I'll get off my soapbox and we can finish the problem. With a bit more algebra, we'll find that the mass of water needed is

$$m_w = \frac{c_{feed} [T_f - T_{i,feed}]}{c_{water} [T_{i,water} - T_f]} m_{feed}$$

We can convert this to liters by remembering that 1 gram of water takes up 1 milliliter of volume. This means that 1 kg of water takes up 1 liter. The volume of water Will needs in liters is the same as the mass of the water in kilograms. Plugging everything in, Will needs 5.40 L of water to make his special feed. To find final weight, add the mass of the water to the mass of the feed and convert the total to pounds. (Remember, there are 2.2 lbs. in a kilogram.) The total weight of the feed mixture will be 64.7 lbs. Better get some help, Will!

---

126 4186 J/(kg C°)

127 Answer: the heat energy lost by the water as it cools from $T_{i,\ water}$ to T_f is gained by the feed as its warms rises from $T_{i,\ feed}$ to T_f. The changes in heat are equal.

Momentum

Momentum: Newton's Third Law

We've tackled forces and run down energy. Now it's time to face our biggest challenge yet – the 350lb defensive lineman of physics also known as **momentum**. It's the only thing between you and the end zone, so it's time to bring out your A-game. If you do, I promise to stop with the sports metaphors.

Problems involving momentum, however, can be some of the most complex problems you'll see at this level. There are a lot of moving parts and a lot of details to keep track of. Imagine gluing two forces problems together using an energy problem, and you've grasped the level of complexity. Having said that, there's no reason you can't master solving these types of problems as well. The conceptual scaffold will help organize your thoughts. Like force problems and energy problems, patience and stubbornness will see you through. Before we get into all of that, however, we should answer one vital question:

What the heck *is* momentum?

Newton's Third Law

It's time for a thought experiment! One of my favorite demonstrations involves office chairs – the kind that have wheels on the bottom so that they can roll. Start by imagining that you're sitting in one of those chairs, near a wall. You put your feet up against the wall and push hard. What would you expect to happen? If you said, "I (and presumably the chair) will roll away from the wall," give yourself a gold star[128]. You can actually try this experiment. Assuming that the chair can roll freely on the floor

128 If you said, "I will experience a force that accelerates me away from the wall," give yourself *two* gold stars instead.

(and that you don't end up tipping the chair), you will find that you do indeed go sailing across the floor. Hopefully this is not unexpected.

However, let's switch things up a little bit. Instead of pushing against a wall, imagine you're pushing against a chair that's identical to yours. Sitting in the chair is the biggest defensive lineman in the NFL, and sitting on his lap is the second biggest defensive lineman in the NFL[129]. What do you imagine happening because of your push? (Don't worry; in your imagination they won't get mad at you for pushing off of them... just as they won't get mad at me for stacking them up like a pair of inseparable teenage girls.) If you could actually perform this experiment, you might be surprised by the result. Just like in the wall example, you and your chair would go rolling away. However, the other chair (and by extension the two football players) would also roll, albeit more slowly.

If you reflect on what you've learned in the section on forces, this should make perfect sense. When you push on their chair, you're exerting a force on them. Unlike the wall, there are no forces acting against your push. Because there's a net force sideways, they have to accelerate sideways. In other words, they have to start moving sideways because you pushed them sideways. You might be tempted to say something like, "But they're so heavy! They must be hard to move!" Think about which direction the gravitational force pulls. (Here's a hint: down.) Can a force pointing straight down cancel or add with a force that's perfectly horizontal[130]? If you exert a net force, there *will* be a change in motion.

What about you? Whether you pushed off the wall or the two football players, you ended up rolling away. This implies that there was a net force exerted upon you as well! "Well, of course," you might reply. "I pushed with my legs!" That's great, until you consider what happens if you sit in the chair and try to push against an imaginary chair full of imaginary defensive linemen. As you might expect, you won't end up moving much, if at all. We talked about the difference between internal and external forces;

129 Hey, it could happen! Besides, if this mental image didn't make you giggle at least in the little bit, you need to go do a Google image search on "NFL defensive linemen" and imagine that situation again.

130 If you can't answer this question, you really need to go back and study the forces section in more depth.

while you can use your legs to push other things, you can't use your legs to push you! That may sound like the most preposterous thing you've ever read. After all, every day people walk, run, stand, jump, and kick with their legs. If their legs aren't pushing them, then what is?

To answer that question we're going to return to our good friend Wile E. Isaac Newton[131]. We've already seen the first two (of three) principles of motion that bear his name; now we're ready to see the third.

Every object that experiences a force exerts an equal force in the opposite direction on the source of the original force.[132]

This means that when you push an object, it pushes you back! This is the force that ends up changing your motion. When your legs push off that chair with the linemen, you end up accelerating *them*. However, at the exact same time they're pushing back against *you*, and it is *that* force – the one that they exert on you – that causes you to move away. When you jump, you push against the ground; in the same instant, the ground exerts an equal force back up on you. The force you exert on the Earth changes the Earth's motion (negligibly, but still). The same amount of force exerted by the Earth back on you flings you into the air – a daring maneuver many people refer to as jumping. Even car tires push against the road, and the road pushes back to propel the car forward. All objects (including people and animals) that appear to propel themselves actually pull or push other objects; it's the "equal force in the opposite direction" described by Newton's third law that actually changes their motion.

But let's go back to the thought experiment with the rolling chairs. Imagine that instead of two big men, the other chair contains a small child.

131 I would say that he's a "super genius" but (to my knowledge) he never tried catching roadrunners with or without questionable mail-ordered devices.

132 A more common (and shorter) way of saying this is "For every action there is an equal and opposite reaction." I hate this version because, as Wolfgang Pauli might say, "it's not even wrong." What counts as an action? Who is the recipient of the reaction? If I hit my brother, is the equal and opposite reaction for him to hit me back, or for him to give me a hug and forgive me? The entire statement feels imprecise and "squishy" to me, and I don't think it helps students build intuition or understanding. If you work with the first statement, the wordy and precise one, you are much less likely to go wrong.

When you bring your feet up to push on the other chair, what will happen?

It should be obvious that the other chair will move much more than yours will. The small child's chair will end up sailing across the room whereas your chair will roll backwards more slowly. The lighter (or, more accurately, less massive) the occupant(s) of the other chair is (relative to you), the faster their chair will move (as compared to yours). Conversely, the more massive the occupants of the other chair (like the football players), the slower their chair will move compared to yours. The speeds of the two chairs after separation are directly related to the masses of the occupants of the chairs. Based on this, physicists have developed the idea of **momentum**. (The plural of momentum, by the way, is momenta.)

Like mass and energy, momentum is a rather difficult thing to pin down. However, there's a simple statement I like to use:

Momentum describes an object in motion.

Back when we first talked about motion, I didn't specify that there should be an object included. Big omission, right? I mean, think about what it would mean to have motion without an object that's moving! I don't know about you, but my mind boggles if I even attempt to make sense of that. The concept of momentum fixes that omission by making the object's existence just as important as the motion itself. In fact, as you dig deeper and learn more about physics, you will eventually come to find that momentum is more basic than either mass or motion. The only fundamental quantity that really matters is an object's momentum. (It's location matters as well, but mostly because where the object is determines what other objects it interacts with, by what means it interacts with them, and how much it interacts.)

The previous definition corresponds directly to the equation for momentum:

$$\vec{p} = m\vec{v}$$

The generally agreed-upon symbol for momentum is p. I leave it as an exercise for the reader to discover why[133]. m as always represents mass; because mass is associated with objects it also represents the presence of an object. The motion (a.k.a. the velocity) shows up in that final v. Remember, the little arrows over the variables are there to remind you that they're vectors – the direction of the motion (and thus, the momentum) matters! The equal sign means equal, of course; another way to read it, however, is "is determined by." It's the verb of the equation's sentence. Translating the math into English, the momentum equation states

(direction matters) **(direction matters)**

(momentum) (is determined by) (object's mass) (object's motion)

Math is an incredibly precise shorthand way of communicating ideas. This is one reason it's used so much in the sciences. Learning to interpret the equations based upon your scientific knowledge is one of the most powerful tools you can master. It's just like learning another language, and like learning German, Latin, or Greek, it will broaden the ways in which you can process and communicate ideas[134]. To that end, compare that to the definition I gave for momentum on the previous page. Can you see how the two are related?

---

133 Whenever you see, "I leave it as an exercise for the reader to..." it usually means either (a) I can't do it, (b) I did it once but I'm not sure it's right so I'm not going to potentially embarrass myself, or (c) I don't want to be bothered with it. In this case, it's definitely reason (b). I found an answer, but I have no idea as to whether or not it's right.

134 Languages, including math, merely provide useful representations for ideas; the word 'apple', for example, is no more edible than 'pomme'. Without it, however, you wouldn't have a mental handle for the juicy white-fleshed red-skinned objects that grows on trees in the fall and are delicious and (somewhat) nutritious. See the appendix "Philosophical Musings and Geometry Reminders" on page 239 for more of my thoughts on the links between math and mental representations.

Newton's Laws and Momentum

Because momentum is based upon motion, and Newton's first two laws dealt with motion, we should review the possible implications of Newton's laws on momentum. Let's start with Newton's first law:

An object's change in motion results

from a net external force.

If an object's motion changes, then its momentum must also have changed. Since changes in motion result from net external forces, then changes in momentum must result from net external forces as well! Thus another way of stating Newton's first law is

Changes in an object's momentum result

from a net external force.

Newton's second law was literally the equation version of Newton's first law. We can do the same thing here, as long as we remember that all changes require some interval of time:

$$\frac{\Delta \vec{p}}{\Delta t} = \sum \vec{F}_{ext}$$

In other words, the change in an object's momentum over time is determined by the net force acting upon that object. (Make sure you understand how I wrote that last sentence based upon the equation above!)

There's an implication to this equation, however. Remember, momentum describes an object in motion. If the momentum changes over time, that could mean that the object's motion changes. This is the situation described back in the chapter on forces, and, to be honest, isn't that interesting – we can already solve those kinds of problems. However, a change in momentum over time could also imply that the object changes, or more likely, that both the object and the object's motion change. It's possible for the object's mass to increase or decrease.

Changing the mass of an object, however, isn't as simple as turning a dial – a fact that is frustratingly obvious to anyone who's ever tried to get rid of the extra mass that always seems to accumulate over the month of December. For our purposes, mass is a conserved quantity – we can't make stuff vanish into nothingness or appear out of nothingness.[135] This means that if an object's mass changes, it's because it's either somehow collecting stuff in from, or tossing stuff out to, its surroundings.

But if I'm bringing stuff in or tossing stuff out, couldn't I consider that stuff to be a separate object? (Hint: the answer is YES!) As an object, this stuff has mass and motion, which means it has its own momentum. And just as I could look at my situation as the original object experiencing a force by gaining or losing this other stuff, I could also look at the situation as the other stuff gaining or losing the original object. In other words, the changes I'm looking at can be described as a change in momentum of the original object due to forces acting upon it from the other stuff, OR as a change in momentum of the other stuff due to forces acting upon it from the original object.

Let's kick this out of the theoretical for a moment and go back to some thought experiments. Remember those neat rolling chairs we were working with at the beginning of the chapter? Imagine sitting in one of those chairs, hanging out and having a good time when suddenly someone tosses a bowling ball to you. Let's imagine that you catch and hold it without injuring yourself. What happens?

---

135 Einstein's theory of special relativity (SR) does establish an equivalence of mass and energy, and predicts that energy can be converted to mass and vice versa. However, notice that even in these conversions the stuff doesn't appear out of nothing or vanish into nothing. It just expands the concept of conservation of energy to include mass as an additional type of energy. As a completely off-topic aside, Einstein went on to develop the theory of general relativity (GR). Sometime later I think he also came up with the theories of admiral relativity, colonel relativity, major relativity, lieutenant relativity, and sergeant relativity. There was even a private relativity at one time, but it collided with public relativity and the two underwent mutual annihilation, producing super-symmetric M-theory. (Or was it super-heroic M-theory?)

Even a slow-moving bowling ball will send you and your chair rolling. The bowling ball exerts a force on you as you catch it; that force will cause you (and the chair you're sitting in) to accelerate. You speed up in the direction the bowling ball pushes you. By the same token, however, you're pushing back on the bowling ball. The force you exert on the bowling ball accelerates it, causing it to slow down. Eventually, your increasing speed and the bowling ball's decreasing speed will match. At that point, you're basically holding onto the bowling ball as the two[136] of you roll along at the same speed. For all intents and purposes, you and the bowling ball have become one object with a larger mass.

Alternatively, let's imagine you're sitting in the chair in the center of the room with a bowling ball in your lap[137]. You lift the bowling ball and toss it away from you. What happens?

This situation is identical to the situations we discussed at the beginning of the chapter. The bowling ball will go flying one way, and you will end up rolling in the opposite direction. The force you exerted on the bowling ball sends it flying off in one direction; the equal and opposite force exerted by the bowling ball on you will accelerate you in the opposite direction. From a certain point of view, your mass also decreased because the bowling ball carried some of the mass away with it.

These two situations encompass what all changes in mass look like – either two (or more) objects combine to form one bigger one, or one object separates into two (or more) objects. The forces exerted by the pieces on each other are what cause the changes in how the pieces move. Here's where it gets interesting, however. Newton's first law states that the change in an object's momentum over time is determined by the net force acting upon that object. However, at the moment of separation or combination, the only force that really matters is the one related to the separation or combination – the force from the other object.

---

136 Strictly speaking there are three objects here – you, the bowling ball, and the chair you're sitting in. However for these examples I'm including the chair as part of "you" since we're not imagining you getting up from the chair.

137 If you can imagine a common set of circumstances that would lead to this situation in real life I'd love to hear it, especially if it doesn't involve setting up a demonstration and/or a physics educator!

Ah, and now we're back to Newton's third law! The first piece (the original object) exerts a force on the second piece (the other stuff). The second piece must exert an equal force in the opposite direction on the first piece. The change in momentum over time of the two pieces is the same, but in the opposite direction! Not only that, remember that forces require contact. By definition, the two pieces are in contact for the same amount of time. (If you don't believe me, touch a nearby surface. Try to stay in contact with that surface longer than the surface is in contact with you. Try not to have your head explode as you contemplate what that might look like.) The forces are the same strength (but in opposite directions); the time interval for the change in momentum is identical. Any changes in momentum of one piece must be matched by an equal but opposite change in momentum of the other piece. Another way of saying this is that changes in the momentum of one object are canceled out by changes in the momentum of another object.

Think back to the bowling ball examples. When you caught the ball, your motion (and thus momentum) increased in the same direction the ball was traveling. The ball, however, had to slow down. As it did so, its momentum decreased. Your momentum went up, the ball's momentum went down, and the total momentum between you and the ball stayed the same. Similarly, when you threw the bowling ball, your motion increased in one direction. The bowling ball's motion increased in the opposite direction. Your momentum increased one way, and the bowling ball's momentum increased by an equal amount in the opposite direction. Again, the total amount of momentum stayed the same.

This result actually holds over all such interactions. When two objects interact, the momentum of the individual objects may change but the total momentum remains constant. This result is known as **conservation of momentum**, and is a powerful tool for solving physics problems. The most straightforward problems we can solve using conservation of momentum problems are collisions. We'll talk more about those in the next chapter.

Internal vs. External Forces, Revisited

However, before we go on, there are two more things I want to say about conservation of momentum. First, I'd like to discuss internal and external forces again. Back when we talked about Newton's first and second laws, one of the things I mentioned was that only external forces count toward changing the motion of an object. But what counts as an internal force? The previous description I gave was basically "any force that's completely inside the object." With the introduction of Newton's third law, we can talk about what that means.

Imagine you're back in that office chair with the bowling ball in your lap. You lift the bowling ball and heave, but someone's covered the ball with Super Glue® – you can't let go! What happens?

Assuming that you don't fall out of the chair, the answer is "not much." As you attempt to throw the bowling ball, you have the beginnings of a collision. The bowling ball gains momentum in one direction, so you gain momentum backwards. You might even roll backwards a little bit as the ball moves away from you a little bit. However, that all changes once your arms are fully extended. Because of the superglue, you can't let go of the ball. Any motion or momentum the ball had will be countered by the force of your glued hands as they pull back on the ball. By Newton's third law, that results in a reaction force on you that points in the opposite direction. That reaction force kills whatever motion or momentum you had gained. The net effect is that although you exerted a force on the ball to throw it away, that force was canceled by the force you exerted on the ball because you were stuck to it. Similarly, the reaction force from the ball on you that would have gotten you moving was canceled by the reaction force exerted upon you by the ball because you were stuck to it. Because you couldn't break into separate parts, all of these forces are internal forces. In effect, an internal force is any force that's inside of an object that does not cause the object to break into pieces. If the object breaks into pieces, then those forces that were internal to the original object become external to each of the pieces.

Newton's third law doesn't just give us some information about how one object reacts to a force from another object, it sets very specific limits on the circumstances under which forces "count." Basically, it means that forces between different parts of an object that don't break the object can't change how the object moves.

Momentum and Fields

The second thing I want to point out concerns a comment I made when I introduced the idea of fields:

"All of the concepts we will discuss – forces, energy, and momentum – apply as much to fields as they do to the more mundane stuff we'll talk about."

When we apply this concept to conservation of momentum, some really awesome ideas pop out. For example, the Earth pulls on the gravitational field. The field then pulls on a falling rock (for example). By Newton's third law, therefore, the rock must pull back upon the gravitational field, and the gravitational field must pull back upon Earth. Just as Earth yanks the rock downward (through the gravitational field), the rock yanks Earth upward (through the gravitational field)! Earth's momentum increases upwards in the exact same amount as the rock's momentum increased downwards. Unfortunately, because Earth's mass is so unbelievably huge[138] it is generally impossible to measure the resulting change in Earth's motion.

But let's replace "Earth" with "the Sun" and "rock" with "Jupiter." Suddenly we get measurable results. As Jupiter orbits the Sun its momentum is constantly changing. Again, according to Newton's third law, the Sun's momentum must also change to perfectly balance the changes in Jupiter's momentum. The Sun is only about a thousand times as massive as Jupiter, which means that it's actually possible to measure the changes in the Sun's motion due to these changes of momentum! This is an effect that we can measure with relative ease. In fact, this method is so easy that it's been used to detect planets orbiting other stars! As such, as a planet orbits its

138 For illustrative purposes, by mass Earth is about one hundred trillion trillion times bigger than you. Written out as a number, Earth's mass is approximately 100,000,000,000,000,000,000,000x your mass. Any changes in momentum that you could possibly produce are insignificant compared to the power of the Dark Side – err, I mean compared to the size of the planet.

star its momentum is constantly changing (just like Jupiter); by Newton's third law, the star's momentum has to change in perfect balance. Over the course of a single orbit, the star will wobble first one way, then the other. By measuring the size of the wobble and how long it takes, we can calculate all of the orbital dynamics for the planet, even if we can't see it directly!

Let's go further, however. Replace "Sun" with "neutron star" and "Jupiter" with "another neutron star." This kind of system is called a binary neutron star. What's interesting about these systems is that as the two neutron stars orbit each other, they strongly affect the gravitational field around them. This should create ripples in the field (so-called "gravitational waves") that travel outward. These ripples carry energy and momentum away from the stars through the gravitational field. Even now, astronomers are working hard to detect these waves. Their properties will tell us a lot, not just about binary neutron stars, but also about the gravitational field that connects us all.

You might say to yourself, "If these waves have never been detected, how do we know they exist?" There are two pieces to the answer. First, similar waves have been measured in the electromagnetic field – light. A similar configuration of charges moving around each other will produce similar ripples in the electromagnetic field. This can and has been measured. Secondly, and perhaps more importantly, it has been observed that binary neutron stars slowly spiral toward each other. As they do this, they move more slowly, losing both momentum and energy[139]. The energy and momentum has to go somewhere. Because the mass of each star is so high it's logical that most of that energy is transferred to the gravitational field. (There are more convincing arguments that lead this conclusion as well; I'm only presenting the simplest.) Based upon our understanding of the electromagnetic field, this energy and momentum should manifest in the gravitational field as ripples (or waves) that we should eventually be able to measure. Stay tuned for more information!

---

139 One of the side effects of this loss of energy and momentum is that eventually the two neutron stars will collide. When they do, massive amounts of energy will be released in a cataclysmic explosion of high energy particles, light, and gravitational waves. Thankfully there are no such systems anywhere near Earth as this energy could easily scour all life from the surface of our planet.

Momentum: Collisions

If you think back over the last chapter, it may become obvious that all of the things we've talked about involved two things interacting, in one way or another. In fact, in building up toward the very concept of conservation of momentum, we focused on two specific types: one thing breaking apart into two things, and two things coming together to become one thing. This applies more generally, however, as *any* interaction between two objects will be described by conservation of momentum. (Although conservation of momentum holds even when there are more than two objects interacting, each additional object increases the complexity of the problem without adding much in the way of new physics. For this reason, it's rare to see introductory problems that involve more than two[140] objects.)

In order to explore the possibilities, I'd like you to imagine a movie about a game of pool:

> The scene opens with a view of a pool table. Scattered randomly on the table are five balls – one that is white, one that is solid red, one that is white with a green stripe, one that is white with a red stripe, and one that is solid black. Two players study the table, pool sticks in hand. The first player – he's dressed in green and gold, so we'll call him Green – steps up to the table. He eyes the red ball, then sets himself up behind the white ball. THWACK! The white ball shoots across the table and rolls right past the red one without touching it. The other player, a woman wearing blue, snickers.

140 If you get a problem with three objects then your instructor probably doesn't like you; if your problems involve four or more, he or she definitely doesn't like you. Thankfully you can still use the conceptual scaffold; assuming the problem is solvable, you should be able to solve it, as long as you're careful.

Blue confidently strolls around to where the white ball has stopped. With a glance, she lines up her shot. POW! The white ball collides solidly into the ball with the green stripe, and the two of them roll away from each other. The ball with the green stripe falls into a pocket with a satisfying thud. Smirking, she moves to line up her next shot while Green looks on nervously. Suddenly, there's a noise off screen, and Blue stands, looking for the source of the commotion. Speedily, Green snatches the ball with the red stripe from the table and replaces it with one that looks identical. Glancing at Blue to make sure she's still distracted, he then grabs the solid red ball and replaces it. He slips the two stolen balls behind his back just as Blue turns back around. For a moment, Blue looks at him curiously; then, she turns to the table to line up her shot. Green smiles back at her with pure sincerity, then watches the table in anticipation as Blue expertly sends the white ball sailing.

In slow motion, the white ball collides with the striped ball that Green replaced. From the look on Blue's face, it's obvious the shot was perfect. However, instead of bouncing its way into a pocket, the ball with the red stripe deforms like a lump of clay, and wraps itself firmly around the white ball. Blue's eyebrows shoot upward in surprise. The two balls, now firmly connected, continue moving slowly across the table toward the black ball. Although they barely tap into it, the collision is enough to set the black ball rolling into a pocket. Blue starts to whip around toward Green, anger blossoming across her face, when suddenly the one remaining ball, the solid red one, begins hissing and shaking. Blue glances at it and frowns. Green, on the other hand, drops to the floor. The underside of the red ball cracks open, and suddenly hot gas shoots forth. The red ball lifts off from the table like a rocket, leaving behind a stream of dust and gas. In horror, Blue's eyes track it as it streaks upward. It strikes the ceiling and, for a split second, nothing happens. Then the ball explodes, spraying wet green paint in every direction. Blue's clothes are ruined by the sudden drenching, and she screams in fury. Cackling maniacally, Green runs off wearing a helmet with two enormous horns on the front.

Who could have guessed that the man in green and gold would be such a poor loser?

In this movie, we have four major interactions between the pool balls. In the first, Green shot the white ball past the red one. Both balls keep doing the same thing. In the second interaction, Blue shot the white ball into the ball with the green stripe. Both balls leave the interaction moving in different directions and at different speeds. The third interaction is when the white ball interacts with Green's tricky red stripe ball. In that case, the two balls are stuck together, and move as a single object afterwards. The fourth major interaction, which is not quite as obvious as the others, is when the red ball shoots dust and gas out from under it. Originally, it (and presumably, the dust and gas) sat on the table as a single object. Later, the dust and gas came out moving downward while the rest of the red ball moved upward.

In the first of these interactions, it should be obvious that momentum is conserved. After all, if the objects basically didn't interact, how could the momentum have changed? (If they don't touch, then no force can be exerted. If no force is exerted, then no momentum or energy can be transferred.) The second and third case are what most of us would classify as "collisions" – one object strikes another. The fourth might be considered something like a "rocket." After all, it's the ejection of the gas that propels the red ball upward.

However, imagine for a second that we run the end of the movie backwards:

> The man in green and gold quickly backs into the room. The woman in blue screams, and green paint springs off every surface to converge at the ceiling and form a red pool ball. The red pool ball falls, slowing as it collects upward-moving dust and gas, finally coming to a rest on the table's surface. Nearby, a chunk of white and red moves across the table slowly. Suddenly, part of the chunk stops and forms the ball with the red stripe. To do so, it spits out a white ball, which shoots across the table back toward the woman in blue.

Notice that in our backwards version, it appears that the red ball collides with the gas and dust; the tangled mess spits out a white ball to change the motion of what's left over. It appears that what was a "rocket" type interaction has become a collision, and vice versa.

Now, it might appear that I've created a specious and unreasonable argument. After all, to change the appearance I had to reverse the movie, and in effect, reverse time. However, momentum should be conserved even if I reverse the direction all of the objects are moving. Just because I switched what came before and what came after ,the interaction itself hasn't substantially changed. Besides, I could have gotten the same effect if I'd instead carefully built a duplicate set of the movie where everything moved the opposite way at the start of the film[141]. Time would be moving forward, but the result would be the same – what appeared to be two objects separating looks instead like two objects merging.

In summary, we have only two types of interaction – not interacting at all, or a collision. The first type isn't so interesting, so we'll drop it. What we're left with is a definition:

collision (*noun*) – An interaction via forces that occurs between two objects. The objects may be combined either before or after the interaction, but not both. During collisions, momentum is conserved.

The great thing about defining collisions this way is that it makes it much easier to remember when the momentum concept scaffold applies. All you need to do is ask yourself: "Self, does this problem involve two objects interacting?" If the answer is, "Why yes, Self, yes, it does," then you're going to need the momentum conceptual scaffold.

Elastic Collisions

There are three different types of collisions: one thing splits into two things, two things combine into one thing, or two things interact but remain two things. All of these possibilities obey conservation of momentum, as we've discussed. However, we haven't really included energy in our analysis.

Now, if you said to yourself, "I thought energy was always conserved as well," then good for you. You're right! What's curious about each of these

141 While this might be difficult, under certain conditions it is quite doable. Simpler situations are easier to prepare this way than complex ones.

possible types of collision, however, is where the final energy ends up. For the first possibility (one thing splits into two), the kinetic energy of the two final objects is, by necessity, greater than the kinetic energy of the original object. In other words, to create the split that produces the two objects, energy has to come from some other source – usually an internal energy. Rockets, for example, convert the chemical energy of the fuel into both the kinetic energy of the rocket itself as well as the kinetic energy of the gas and dust that makes up the exhaust. Foam dart guns convert the internal energy of the spring launcher, not only into the kinetic energy of the foam dart, but also the kinetic energy of the gun's recoil (which ends up being absorbed by your hands and arms).

In the second possibility (two things combine into one), the opposite occurs[142]. The kinetic energy of the two initial objects is, again by necessity, less than the kinetic energy of the final, combined object. To force the objects to combine, some of the energy must be shifted into alternate forms. In a car crash, for example, energy is lost in heating, bending, and sometimes tearing of the metal of the cars. Some energy is also lost through sound – the loud "CRASH" that always accompanies the aforementioned heating and bending.

The third possibility (two things interact but remain two things), is perhaps the most interesting. It's certainly possible to imagine a car crash where the two cars remain separate. However, it's also possible to imagine the two cars bouncing off each other with nary a scratch or dent[143]. An example that requires less stretching of the imagination, however, is that of pool balls. Pool balls don't break, stretch, tear, or dent, nor do they contain chemical energy or some other stored energy that could be explosively converted into kinetic energy[144]. In fact, if we could watch the surface of the pool balls during such a collision, we'd probably notice the surface of the ball bend slightly before flexing back, just like a miniature spring. Any kinetic energy that would have been lost is returned by this

142 This shouldn't be a surprise, by the way; remember, combining two things into one is basically the time reversed version of splitting one object into two.

143 Matchbox ® toy cars, for example, often behave this way in crashes.

144 That, however, would make for an exciting game of pool; at any point the balls might combine into a larger pool ball, or split explosively into smaller pool balls... Man, now I really want someone to invent this game.

small push. The kinetic energy of the two pool balls after the collision is approximately the same as it was before the collision. These kinds of collisions, where the only energy transferred is kinetic energy, are called **elastic** collisions.

One thing I think it's important to point out is that even collisions between pool balls aren't really elastic. In a perfect elastic collision, there are no other kinds of energy involved other than kinetic – no potential energies, no internal energies, nothing. In the real world, it's almost impossible to eliminate the production of heat and sound in such collisions. Imagine two pool balls slamming into each other and then rolling away, all without making a single sound! Even in space, where there's no atmosphere to transmit the sound to your ears, the collision will create vibrations within the objects themselves – sound that can't escape. Eventually these vibrations will be converted to heat. Realistically the best you can produce is a practically elastic collision – one in which the amount of energy that gets converted to heat or sound is very small compared to the overall kinetic energy[145].

Inelastic Collisions

By definition, any collision that isn't elastic is an **inelastic** collision. In general, any time an energy other than kinetic energy changes during a collision, it's an inelastic collision. In many collisions two objects interact and separate, but quite a bit of the kinetic energy that the objects had is converted to sound, heat, and other internal energies – like the breaking, bending, and denting of the objects themselves. These collisions are **partially inelastic**, and are by far the most common kind. They're also the most difficult kind of collision to solve mathematically. Because of this, it's rare to see these kinds of problems in an introductory physics course.

145 Even atomic collisions aren't perfectly elastic. When one atom "collides" with another there's a good chance that the structure of the atom will change; energy will go into boosting the orbit of an electron in one (or both) atoms. Even subatomic collisions can be problematic due to the internal structure of protons and neutrons; in addition, it becomes much harder to ignore quantum mechanical effects that challenge our simple definition of "collision".

What's much more common instead are the kinds of collisions I've already described – two objects combine to form one, or one object splits into two. Because these represent the extreme case, they're called **perfectly** inelastic collisions. There are two categories of perfectly inelastic collisions that people tend to find fascinating: rockets and explosives.

Rockets and Explosives

Rockets and explosives have a lot in common. Both convert internal forms of energy into the kinetic energy of the separating pieces. The only difference, really, is that rockets confine and direct the movement of the pieces, whereas in an explosion the pieces are pretty much free to move however they like. This is why rockets that lose confinement often explode, and explosions that are confined can send one or more objects rocketing off at high speed[146].

Fascinatingly, even in these potentially complex situations, conservation of momentum easily describes the result. Rockets, for example, move faster and faster not only because of the energy released by the fuel, but also because the gases and other products are thrown out the back. The fuel provides energy; the ejection of the gases conserves momentum. All rockets, then, need these two things: fuel and something to throw – a.k.a. reaction mass.

For jets, this is not much of an issue. Jets carry fuel to provide energy. They then use the air itself as reaction mass. Air enters the front of the jet, gets heated by the burning fuel in the engines, and comes out the back much faster than it went in. The increase in the air's momentum is canceled by an increase in the jet's momentum in the opposite direction[147]. Rockets, however, are designed to operate in environments where there's no air. This means that not only does the rocket have to carry its fuel with

146 A gun, for example, is basically an explosion confined by a tube and a stopper.

147 The jet, however, loses momentum to the air as it moves forward. The front edge of the jet is constantly experiencing a series of partially inelastic collisions. These serve to limit how much momentum the plane can obtain, which limits how fast the plane can go.

it, it also has to carry its reaction mass! To conserve space, most rockets use whatever is left over from burning the fuel as reaction mass. Chemical rockets, for example, combine hydrogen and oxygen to form water. This reaction releases energy, and the water can then be flung out the back as reaction mass. However, this is quite expensive. Most rockets designed to reach low Earth orbit using this reaction are more than 80 percent fuel by weight.

One staple of science fiction is the ability for a spaceship to increase its momentum without actually throwing anything out – a reaction massless (or "reactionless") rocket. While (strictly speaking) this isn't possible, it's possible to create a rocket that *looks* like it isn't throwing anything out. For example, it is conceivable that at some point we may be able to produce gravitational waves, just as neutron stars do. If so, we can throw those out of the back of a hypothetical ship. The momentum those waves carry away will be balanced by the momentum our ship gains in the opposite direction[148].

Even if we remain in the realm of the immediately feasible, however, it is possible to make "more efficient" rockets. One particular method that really blurs the boundary between rocket and explosive comes to mind. Code named "Project Orion" by the U.S. Government, it was a study dedicated to using fission-based nuclear weapons as the fuel and reaction mass for rockets[149].

The basic design involved putting a nuclear bomb directly behind a big shield attached to the spaceship. When the bomb explodes, the fragments are semi-confined by the shield, and end up gaining momentum mostly in one direction. The spaceship, in turn, gains kinetic energy from the explosion and momentum in the opposite direction from the fragments. A little while later you repeat the process, and your spaceship speeds up a little bit more. Such a ship could quickly and efficiently get up to extremely high speeds, and almost all of the proposed designs are less than 50 percent

148 This concept has already been proposed using a different field; instead of gravitational waves, it uses electromagnetic waves – using light as the reaction mass instead of gravity!

149 You know, I really wish I had a time machine just so I could go back to listen in on some of the conversations around these ideas.

fuel by weight. To be honest, there are some compelling reasons to explore these ideas further.

However, there are a few minor potential drawbacks as well, such as nuclear fallout and contamination, accidentally irradiating the ship's cargo and crew, and even the possibility of partially or totally destroying the ship in a cataclysmic ball of hellishly bright light. Most of the more routine engineering problems are not only solvable, they've been solved. The real issue (for me, at least) is one of risk versus reward: Does this design offer an advantage so great that it overcomes the potential devastation if something goes horrifically wrong?

Right now I don't think I'd willingly ride in such a ship without a *really* good reason (such as the imminent destruction of the entire Earth), but you've got to admit – it would be one heck of a ride!

Momentum: Concept Scaffold

Collision problems – those based upon conservation of momentum – are generally the most complex problems you'll see in introductory physics. These problems are not necessarily harder, but they have a lot of detail, and keeping track of them is vital. The conceptual scaffold for momentum problems is specifically designed to attack this issue:

1. Prepare
2. Draw two pictures
3. Identify the momenta
4. Determine the elasticity of the collision
5. Choose a system of coordinates
6. Break momenta into components
7. Identify initial conditions
8. Identify final conditions
9. Apply conservation of momentum (and...)
10. Do math and solve

The rest of this chapter explains each step in more detail. In the next chapter, I'll show you the conceptual scaffold in action by applying it to several different example problems.

It's been said before, but I'll say it again: The conceptual scaffold is not the solution to the problem. You'll need to do the heavy lifting; the scaffold is only here to help you as you do it. Fortunately, at this point, you have quite a bit of experience under your belt. You've worked with two conceptual scaffolds before. Most of what you'll see here should already be structurally familiar. In addition, collision problems tend to be conceptually straightforward, so keep track of the details and you'll be fine.

Remember as well that every step of the conceptual scaffold is necessary. Each step builds upon what was done before to prepare for what is to come. To successfully complete a given step, you will have had to complete the previous steps correctly. It stands to follow that successfully completing a given step means that you're prepared to successfully complete the steps that lie ahead.

1. Prepare

As I said above, momentum problems are some of the most complex problems you'll see in introductory physics. So it's important to take in a deep breath at this point and get ready for the long haul. Is your pencil as sharp as your intellect? Is your eraser as flexible as your wit? Is your table as unyielding as your will?

Read the question. Let your thoughts steep in the description, and spend some time imagining what's happening. Is one thing splitting into two? Are two things combining into one? Or are two things coming together, colliding, then splitting apart? Play the movie of the collision in your mind until you have a basic idea of how things unfold. Read the question again to pick up anything you missed the first time. If you have questions, make sure you find answers (usually by asking your professor). Stretch, if necessary, and relax. You can do this. You're good enough, you're smart enough, and doggone it, people like you!

Oh, and you should probably release that deep breath. I don't want you to pass out from lack of air.

Looking Forward: Make sure you understand the situation as described by the problem statement. If you're having trouble imagining what's going on, you should get more information. (Remember, asking for clarification is not the same thing as asking for an answer!)

2. Draw two pictures

Since momentum has direction-like forces, we need a clear picture of what is where, just as in force problems. However, because this is a conservation problem, we also need to keep track of what's going on right before the collision as well as right after the collision, as we would in an energy problem. That means you'll need to draw a picture for each of these points in time. The first picture you draw will depict the situation immediately before the collision occurs – the rocket just before the hot gas is released, or the golf ball just before it smacks into the unsuspecting bystander's crotch. This will be the "before" picture. The second – or "after"– picture will show the situation immediately after the collision – the cloud of hot gas and the relieved rocket, or the bystander with the shocked expression clutching the golf ball. While it's perfectly fine to include other stuff, the focus of your drawing should be on the objects experiencing the collision directly – the rocket and the gas, or the golf ball and the bystander.

Please note: It's very important that these pictures occur *immediately* before and after the collision. There shouldn't be any time or opportunity for the pieces to interact with anything else! If you get this piece wrong, one of two things will happen. Either you will discover it in a later step and have to start over completely (which sucks) or you'll get to the end and find that you can't solve the problem (which sucks even more). Do your best to get this right the first time.

<u>Looking Back</u>: Read the problem statement, and make sure that you've used the description provided to draw accurate pictures. All of the key elements of the problem should be represented in your drawing.

<u>Looking Forward</u>: Just like in the forces conceptual scaffold, you'll be adding stuff to the picture as you go. Please make sure the picture is big enough to easily accommodate those additions.

3. Identify the momenta

When doing forces problems, you shouldn't add the motion to your drawing because motion is not force. Here, however, we're going to do the opposite. In each of the pictures, we want to know how the objects are moving. In other words, for each of the objects in the "before" picture, draw in an arrow to show the direction that object is moving just before the collision. Then label the arrow as the object's momentum. Usually I use something like $p_{i,<object>}$; it's a name, however, so you could label the arrows "Phil" and "Dixie" if you really wanted to. Just make sure that you know which object's momentum "Phil" represents! Repeat this for the "after" picture. (For those, I tend to use $p_{f,<object>}$.) If you have an object that isn't moving, you should still mark it. However, instead of an arrow you may want to simply put "p = 0," or something similar. In any case, make sure that you can see at a glance how the objects are moving both before and after they collide.

Looking Back: Reread the problem statement and make sure you haven't missed anything. It should be immediately obvious from your pictures and arrows which objects are colliding and which direction they'll be going in after they do so.

Looking Forward: Look over your momenta and consider their directions. Is there any direction that seems special? Given that this is a collision problem, it is much more likely that the momenta will be lined up in the same direction than it would be with a force problem. Keep these directions in mind, as they may come in handy later on.

4. Determine the elasticity of the collision

Now, based upon what you've done so far, you need to figure out whether the collision is perfectly inelastic, (almost) perfectly elastic, or partially inelastic. Perfectly inelastic problems are both the easiest to identify and the easiest to solve. Any problem where one object splits into two or where two objects combine to form one is a perfectly inelastic problem. For these, you will not need to worry about conservation of energy. If the collision is not perfectly inelastic, your next goal is to figure out whether the collision is perfectly elastic or partially inelastic. It's difficult to give hard and fast rules about figuring this out. However, I can give a few guidelines. First, the problem may say to treat the collision as perfectly elastic or to ignore any energy losses. That's the easiest and clearest outcome. Other indicators are more subtle: for example if it says the objects will clearly bounce without deforming, like pool balls. If you can't tell, you should look for indicators that the collision is partially inelastic, such as the inclusion of energy information[150] or more information about how the objects are moving than appears to be necessary. However, the best way to figure this out is through experience. The more collision problems you solve, the easier it will become to identify the collision type.

For perfectly elastic problems, you will need to use conservation of energy; however, because the energy remains as motion, you will only have kinetic energy terms in your equations. If the problem is partially inelastic, however, you will have to play it by ear. If, for example, you have the energy information, you may need to use conservation of energy; otherwise, you may be able to hobble by with the extra information provided. That said, partially inelastic collision problems tend to be rare in introductory physics courses. If you think you have one, you should ask the professor for clarification. It's always possible that the person who wrote the exam made a mistake or did not write the question clearly enough.[151]

In any case, write down the collision type prominently. You'll refer to it later. (I usually jot it down right between the "before" picture and the "after" picture so I can't miss it.)

Looking Back: Examine the picture to make sure that what you've described here is what is actually happening. In other words, you've chosen the elasticity based upon the problem description. Now make sure that your chosen elasticity matches what you drew in the previous step.

---

150 Examples of energy information are statements like "<Some amount> of <energy source> is burned/consumed/exploded, releasing <so much energy per amount>", or "<some amount> is lost as <some form of energy>".

151 Let me take this opportunity to reiterate: if you have done your best to figure things out and can't, please ask your professor! Asking for (and receiving) help is part of the learning process. Never be ashamed to ask questions, even ones that might seem (at first glance) to be obvious, stupid, or bad.

Looking Forward: The next few steps will require extra work if the collision is perfectly elastic or partially inelastic, so make sure you make this choice correctly. It really sucks to get close to the end, only to realize that you've made a mistake here and need to redo everything.

5. Choose a system of coordinates

With forces problems, choosing a good coordinate system was important; the wrong choice could make a problem hard or even impossible. With momentum problems, on the other hand, the choice of coordinate system is less important. Although your choice can make the problem easier or harder, it's unlikely that it will make the problem impossible. So choose whatever set of coordinates you like.

Still, it's always a good idea to choose a coordinate system that will make the problem easier. Try to line up as many momenta as possible with coordinate axes. Remember, to count as a coordinate system the coordinate axes have to stay at right angles to each other, so this method works particularly well when most of your momenta are also at right angles to each other. You will only have one coordinate system for the entire problem, so consider the directions of the momenta both before and after the collision!

As with forces, the axis labels are purely for reference, so please feel free to be creative.

Looking Back: Review all of the momenta you've labeled and make sure you haven't forgotten any. As with forces, the coordinate system will help you translate your diagrams into equations, so it's important that you've appropriately labeled all of the momenta.

Looking Forward: Again, the purpose of the coordinate system is to make sure that your equations match your pictures and make sense. Choosing an appropriate coordinate system now may make the next steps easier. As long as you follow the rules of coordinate systems, you should be fine!

6. Break momenta into components

Like with forces problems, we need to break our vectors into coordinate components so that we can build our equations[152]. In this case, the vectors – the quantities with direction and size that we represented with arrows – are the momenta. We'll need to use the same tools to break these into components that we used back in the first section to break the forces into components. In other words, cue the calming mantra: **SOHCAHTOA**! This works identically to how it worked before. Just so you have it at your fingertips, however, I'll repeat the definition and quick explanation here:

"SOHCAHTOA is an acronym to help you remember the relationship between the sides of a right triangle and the angles between those sides:

Sine
Opposite
Hypotenuse
Cosine
Adjacent
Hypotenuse
Tangent
Opposite
Adjacent

If you choose one of the smaller angles in a right triangle (remember *Bob*?), the sine of *Bob* is equal to the length of the side directly across from *Bob* divided by the length of the hypotenuse (the longest side of the triangle). The cosine of *Bob* is equal to the length of the side adjacent (or next to) *Bob* divided by the length of the hypotenuse, and the tangent of *Bob* is equal to the length of the opposite side divided by the length of the adjacent side."

In all other ways, this step is identical to "Step 5: Break forces into components" of the forces conceptual scaffold, for both good and for ill.

152 It's almost impossible to do the math without converting to components in one way or another – that's the purpose that the unit vectors **i**, **j**, and **k** serve, after all. And if you don't know what unit vectors are, don't worry about it. Move along, move along. Nothing to see here.

Thankfully, however, figuring out the angles tends to be less onerous here than it can be with forces. This is where working with momentum is much better than working with forces: If you don't like the angles that are showing up, you can always back up a step and choose a different coordinate system without much risk!

Looking Back: Make sure that every momentum you identified lies entirely along one of your chosen coordinate axes or is broken up into components. Remember, you have to do this for all of the momenta in both the "before" and "after" pictures!

Looking Forward: When we build our equations in the next few steps, we'll be looking to see whether the components point in the same direction or opposite direction of the appropriate coordinate axis. That will determine whether that component counts as positive or negative (respectively) in our equations. Look over your drawing. If you can't immediately say whether a given momentum component is positive or negative, you should immediately review your work for mistakes.

7. Identify initial conditions

Now we've gotten to the heart of the matter. Momentum problems, like energy problems, are about change. That's why we had two pictures: They'll help us get a handle on how things start and how things finish. In this step, we'll set up the part of the equation that describes how things started; therefore, we'll focus solely on the "before" picture. In particular, we want to record all of the momentum components for each coordinate direction so that we can eventually calculate the total momentum in each direction before the collision.

For example, suppose that your coordinate axes are "Alice" and "Rabbit." To complete this step, you should build a list of momentum components that point either in the same direction or opposite, labeling them as "Alice." You should then build a separate list of components for "Rabbit." As always, components that point in the same direction as the axis get a positive sign; components that point in the opposite direction get a negative sign. Don't worry about actually calculating out the number.

For now we just want to keep a record of where the momentum starts. Once you have a list of all of the different momenta, add them up for each direction. This will give you the total initial momentum by direction.

There's one last thing we need to consider, however. If the collision was (almost) perfectly elastic, we *definitely* need to worry about conservation of energy; if it was partially inelastic, we *might* need to worry about it. (If the collision is perfectly inelastic, you probably don't have to worry about energy. In that case, skip to Step 8 and advance your piece two spaces on the board.)[153]

Don't worry too much. You already have the tools necessary to account for energies: the conservation of energy scaffold! If your collision is perfectly elastic (or it's partially inelastic and you have the energy information you need), then effectively you will also insert "Step 2: Determine the total initial energy" from the energy conceptual scaffold here. Again, we're just concerned with the "before" picture – the situation immediately before the collision.

One last thing to point out: If you identified your collision as perfectly elastic, then only the kinetic energies should matter. You shouldn't have to worry about any internal energies or potential energies. Remember, by definition, only the kinetic energies change in perfectly elastic collisions. If it really looks like some other kind of energy might be involved, then most likely your "before" picture doesn't represent the situation *immediately* before the collision. The best thing to do in this case is to start over completely. If you've messed up your "before" picture in this manner, there's not much you can do to fix it without risking serious trouble down the line.

Looking Back: Once you've completed this step, every momentum in the "before" picture should show up in at least one of the lists. Make sure you caught everything! (You don't have to put in any momentum that you've recorded as being zero.) If you find a momentum that shows up in your

153 This question of "do I need to include conservation of energy?" is why this step is "Identify initial conditions" instead of "Determine the total initial momentum". You'll always need to determine the total initial momentum, but you may need to do more, depending on the type of collision.

"before" picture but not in your lists, then something's wrong. You'll need to go back to the previous step and make sure you've broken all of your momentum vectors into components appropriately. Make sure as well that you've included the conservation of energy piece if necessary.

Looking Forward: Again, for this step you only need to deal with the "before" picture. The "after" picture will get a step of its own very soon.

8. Identify final conditions

Now that we've found out where all of the momentum (and energy, if necessary) started, we now need to figure out where it ended up. This step is identical to the previous step in all ways except one: You are now focusing on the "after" picture rather than the "before" picture. Make a list of the momentum components for each axis direction, then add them up, just like you did in the previous step. This will give you the total final momentum by direction. (As always, components that point in the same direction as the axis are positive; those pointing in the opposite direction are negative.[154])

Similarly, if you're dealing with conservation of energy, you'll need to insert "Step 3: Determine the total final energy" from the energy scaffold. Again, for a perfectly elastic collision, only the kinetic energies should matter. If it really looks like some other kind of energy might be involved, then most likely your "after" picture doesn't represent the situation **immediately** after the collision, and you'll need to start over completely.

Looking Back: Once you've completed this step, every momentum in the "after" picture should show up in your lists. If any momentum shows up in your "after" picture but not in at least one list, then something's wrong. Go back to the previous step and make sure you've broken all of your

154 I realize that I'm repeating myself here, but I really can't emphasize this enough – determining the sign of the component based upon the direction is a vital part of building your equation. If you do this incorrectly there is no way you can get the right answer! In my experience this can be a problem point for many students. As some great historical figure once said, "Eternal vigilance is the price of good grades!"

momentum vectors into components appropriately. Again, make sure you've included the conservation of energy piece, if you need it.

Looking Forward: Eventually we will want to calculate the values for all of the momenta (and energies, if appropriate). If you know the equations you need, please feel free to write them down. This isn't absolutely necessary, however, and you certainly shouldn't worry about plugging in numbers yet. Save that for Step 10!

9. Apply conservation of momentum (and...)

After a lot of work, we're finally ready to apply the core concept of conservation of momentum. (Whew!) I'd love to tell you to sit back and appreciate the amount of work you've done simply to prepare for this step, but there's still a lot left to do.[155]. You'll definitely need to apply conservation of momentum; you may need to apply conservation of energy as well[156].

The basic idea of conservation of momentum is that all changes in momentum are balanced. The total momentum during a collision or interaction won't change. In other words, the total initial momentum should equal the total final momentum (by direction).

So let's do that. For each direction, set the total initial momentum equal to the final total momentum. Let's return to the "Alice" and "Rabbit" example. You should have a sum of momenta in the "Alice" direction from Step 7 (the "before" case). That will equal the sum of momenta in the "Alice" direction from Step 8 (the "after" case). Similarly, the sum of "before" momenta in the "Rabbit" direction from Step 7 will equal the sum of the "after" momentum in the "Rabbit" direction from Step 8. If you had to calculate energy, then this is where you'll insert "Step 4: Apply conservation of energy" from the energy scaffold as well. You'll set the en-

---

155 Actually, go ahead and take a moment to pat yourself on the back. You deserve it for getting this far.

156 This is the "(and...)" from the step title; you may or may not need to worry about the conservation of energy here.

ergy just before the collision (from Step 7) equal to the energy immediately after the collision (from Step 8).

At this point, you should have one (best case) to three (worst case[157]) equations. You're now ready for the last step!

Looking Back: Conservation of momentum problems require keeping track of a lot of details. As long as you've done this carefully and neatly, you should be ready to move forward.

Looking Forward: If you don't have easy access to everything you need to calculate the energies or momenta (for example, any equations or constants that might apply or that you may need to look up), now is the time to grab them. You'll need them to solve the problem!

10. Do math and solve

As with all of the conceptual scaffolds, at this point you're done with the physics. If you only have one equation, congratulations! Solve for the unknown variable and use that to answer the question.

More likely, however, you'll have two (or maybe even three) equations with multiple unknown variables. For these, you'll have to whip out your 1337[158] algebra skills. Whether you call it "solving a system of equations," "variable isolation and substitution," or "a giant pain in the butt," you'll be doing the same thing – calculating out what each of the unknown variables must be based upon the equations. There are a variety of techniques to do this; unfortunately, I can't generally predict which will work best or require the least work. Doing practice problems will build your intuition and expertise. Remember, the only way to get better is by doing. In any case, the one suggestion I will make is to write down every line of your calculations. Changing too many things in a single step makes it difficult to hunt down problems when (not if) they occur.

157 If you have more than three equations coming out then you probably could have chosen a better coordinate system. Don't worry, however; the problem is probably still solvable. Probably.

158 900D j0B 1Ph j00Z (4|\| r34D 7|-|1$!

Looking Back: If at any time you are unsure of where a particular term in your equation(s) came from, you should be able to track backwards to identify its source. This is the strength of being careful and systematic. Everything should be linked both forward and backward in a logical way. If something comes out differently than you expect, take the time to try to figure out why. Does your answer make sense given the problem statement and the type of collision?

Congratulations! You've done it!

Momentum: Examples

Here you will find a few sample problems based on the momentum conceptual scaffold. I've set the problem statements here so that you can first try applying the conceptual scaffold, then check your work against the example solutions.

Example 1: A half-loaded boxcar (m = 75,200 kg) breaks free from a train engine (m =136,000 kg) and ends up rolling along the track at a constant speed of 10.6 m/s. The train's engineer, Simo Muinonen, sends the engine in pursuit of the boxcar at a constant speed of 22.5 m/s. As they collide, Simo deftly inserts a coupling pin between them to lock the two together, then pulls the brakes, bringing both to a stop. How fast are the train and boxcar moving right after Simo inserts the coupling pin?

Example 2: Jonathan "the Birdman" Sanok, the daring starfighter pilot, is flying his S.P.A.R.R.O.W. high-performance starfighter (m = 13,600 kg) at 1420 m/s when he realizes that he's on a collision course with an asteroid. To avoid the asteroid, Jonathan fires the S.P.A.R.R.O.W.'s maneuvering thrusters for eight seconds, releasing 5.75 kg/sec of reaction mass at an exhaust speed of 63,500 m/s. The released gas moves away from the ship at a right angle to its original motion. This adjustment allows "the Birdman" to evade the asteroid; however, it also puts him off course. How fast and in what direction is the S.P.A.R.R.O.W. now moving (relative to its original motion)? Is he significantly off course? Explain your answer.

Example 3: Ormond Harriott is playing marbles with her granddaughter, Dasia. Dasia fires her marble (m = 3.18 g) at 3.77 cm/s, and it collides head on with Ormond's stationary marble (m = 4.25 g). (a) How fast is Ormond's marble moving just after the collision? (b) How fast is Dasia's marble moving just after the collision?

Example 1

A half-loaded boxcar (m = 75,200 kg) breaks free from a train engine (m =136,000 kg) and ends up rolling along the track at a constant speed of 10.6 m/s. The train's engineer, Simo Muinonen, sends the engine in pursuit of the boxcar at a constant speed of 22.5 m/s. As they collide, Simo deftly inserts a coupling pin between them to lock the two together, then pulls the brakes, bringing both to a stop. How fast are the train and boxcar moving right after Simo inserts the coupling pin?

1. Prepare

Preparing to do collision problems is a little different than preparing for energy problems, or even force problems. Because these problems have so many details, they often take longer to solve and have many more pieces to keep track of. Because of this, I suggest that in addition to your writing materials, you also make sure you have snacks and a comfy seat. There's nothing worse than having your butt fall asleep halfway through your work because your chair is too hard!

Regardless of the type of problem, however, you must still make sure that you understand what's happening in the problem. Take a moment, sit back, and close your eyes – wait, not yet! Finish reading this paragraph first! Okay, when you close your eyes, can you imagine the train? Can you see the boxcar rolling down the track, with the engine in hot pursuit? The smell of oil and steel tickling your nose as Simo's arm extends, coupling pin in hand, ready to corral the wayward boxcar? Can you see the beads of sweat rolling down his brow as concentration and care carve his face? Can you hear the clear metallic clang as the pin slides perfectly into place and the screech of metal as the two lock together? I can! If you can too, open your eyes. We're ready to start!

2. Draw two pictures

Remember, momentum problems are like energy problems – we're going to be considering an initial situation and a final situation. In this case, the initial situation is right before the collision, and the final situation is right after the collision. In order to account for all of the momenta, we need pictures! That means that for this problem, we'll need to draw two pictures: one that describes the scene immediately before the collision, and one that describes the scene immediately after. It's important that our images show the situation as close to the collision as we can get (both before and after). We want to eliminate as much outside influence as we can. (In other words, it doesn't do us any good to draw the train and boxcar stopped in the "after" picture, as a lot of extra stuff[159] has to happen to get to that point!)

So, here's a quick depiction of the situation just before the collision:

and one just after:

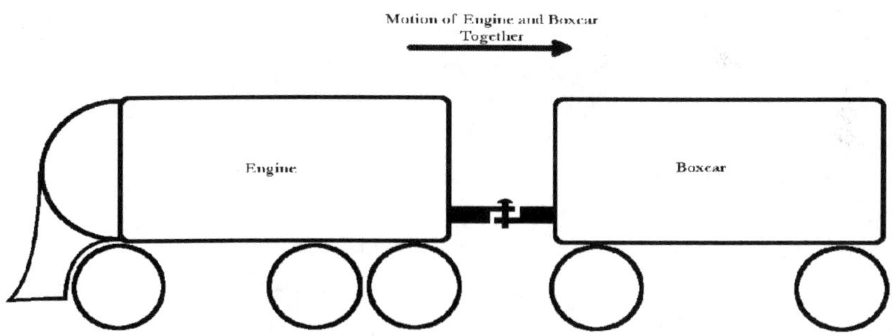

I've tried making these drawings slightly nicer than I have in the past. By the way, that's supposed to look like a diesel engine, not a bullet with a wheels and a beard[160]. Yeah, even I make fun of my drawings.

159 "Extra", in this case, means "in addition to the collision".

160 Although, if I turn my head just right, that beard looks a little bit like butterfly wings....

3. Identify the Momenta

In the above pictures, I've marked the direction that each object is moving, both before and after the collision. Before the collision, the boxcar is moving to the right, and the train engine (and Simo) are moving to the right. Both of these, then, have momentum rightward.

When I watch the film of this situation in my mind's eye, everything is still moving to the right immediately after the collision. However, once the engine and boxcar are connected, they have to be moving together. After the collision, then, they've effectively become a single object, and have a shared momentum and shared speed.

4. Determine the elasticity of the collision

Now it's time to determine if this collision is perfectly (practically) elastic or perfectly inelastic. It could also be partially inelastic, but those situations are rare in introductory physics. In this case, we have two objects that collide and become one object. If we review the types of collision, it's pretty clear that this is a perfectly inelastic collision. That means we don't have to worry about the kinetic energy, as the final total energy includes things like noise, heat, and scratches on the metal.

This step may seem easy — even trivial. However, it's incredibly important. As you'll see in Example 3, determining the type of collision here has major consequences going forward.

5. Choose a system of coordinates

If we glance back at Step 3, we'll notice that everything seems to be moving rightward both before and after the collision. Let's make that direction one of our chosen axes. We'll call that “Used.” Our other coordinate direction doesn't matter much. It needs to be at right angles to “Used,” so let's choose upward. We won't be using it, however, so we'll name it “Not”!

Updating the picture to include our coordinate axes, we get

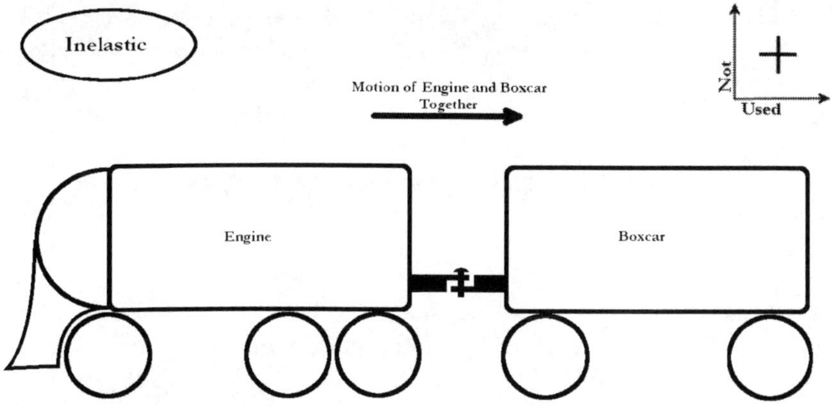

Even though I've only shown the "After" picture, we'll use the same coordinate system for the "Before" picture, too.

6. Break momenta into components

If we've chosen our coordinate system wisely, this step will be easy. Everything points in the "Used" direction[161]! Let's write them out carefully, just to get practice:

	Before:	*After:*
"Used":	$+\mathbf{p}_{\text{engine}}$, $+\mathbf{p}_{\text{boxcar}}$	$+\mathbf{p}_{\text{together}}$
"Not":	*0, nada, nothing, zip.*	*0, nada, nothing, zip.*

This is about the easiest that Step 6 can be. As you'll see in Examples 2 and 3, this step can be much more complex. For now, we'll take it and run!

161 Don't feel too bad about how "Used" this direction is. It trained for this.

7. Identify initial conditions

Due to our work in the previous steps, this is trivial. Because we're dealing with an inelastic collision (from Step 4), we only have to deal with conservation of momentum. So let's write out our total initial momentum by direction carefully:

"Used": $\mathbf{p}_{\mathbf{engine}} + \mathbf{p}_{\mathbf{boxcar}} = \mathbf{m}_{\mathbf{engine}}\,\mathbf{v}_{\mathbf{engine}} + \mathbf{m}_{\mathbf{boxcar}}\,\mathbf{v}_{\mathbf{boxcar}}$

"Not": 0, nothing, nada, zip

Just to add some heft to this step, I've gone ahead and expanded the equations for the momenta above. If you'd prefer to leave that for Step 10, that's awesome, too!

8. Identify final conditions

This step is also pretty easy, as we've done all of our previous steps carefully. Our final total momentum, by direction, is:

"Used": $+\mathbf{p}_{\mathbf{together}} = \mathbf{m}_{\mathbf{together}}\,\mathbf{v}_{\mathbf{together}}$

"Not": 0, nothing, nada, zip.

Hopefully this step, as well as the last, contained no surprises. If you can't see how I got this, please review the table in Step 6.

9. Apply conservation of momentum (and...)

Conservation of <something> means that we can't gain or lose <something>, we can only move it around. So, conservation of momentum states that we can't create any new momentum or destroy momentum. Anything we started with should still be around when we end. In other words, our initial total momentum should be equal to our final moment for each direction. I've dropped the "Not" direction here because it's "Not" really interesting.

$$m_{engine}v_{engine} + m_{boxcar}v_{boxcar} = m_{together}v_{together}$$

The "(and...)" in this step is purely for elastic collisions. If this had been an elastic collision, we'd have to worry about the kinetic energies as well. It's not, so there's no "(and...)." We're done!

10. Do math and solve

We've used physics principles to craft our equation(s). Now we can use math principles to solve our equation(s). This problem should be fairly straightforward; however, there is one potential tripping point. You must remember that the mass of the train and boxcar together is the same as the sum of the mass of the train and the mass of the boxcar. A little bit of algebra gets us

$$v_{together} = \frac{m_{engine} v_{engine} + m_{boxcar} v_{boxcar}}{m_{engine} + m_{boxcar}}$$

and we can plug in our values at this point to find the final speed of the train after the collision.

$$V_{together} = 18.3 \text{ m/s}$$

For comparison sake, that's more than 40 mph (or 65 kph)[162]. It may take Simo a minute or two to bring that thing to a stop!

This problem may have seemed fairly easy. That's the point of the conceptual scaffold: A systematic reductionist approach to solving problems produces a straightforward and reproducible solution.

162 It's also important to point out that this speed isn't simply the average of the train's speed and the boxcar's speed before the collisions, either.

Example 2

Jonathan "the Birdman" Sanok, the daring starfighter pilot, is flying his S.P.A.R.R.O.W. high-performance starfighter (m = 13,600 kg) at 1420 m/s when he realizes that he's on a collision course with an asteroid. To avoid the asteroid, Jonathan fires the S.P.A.R.R.O.W.'s maneuvering thrusters for eight seconds, releasing 5.75 kg/sec of reaction mass at an exhaust speed of 63,500 m/s. The released gas moves away from the ship at a right angle to its original motion. This adjustment allows "the Birdman" to evade the asteroid; however, it also puts him off course. How fast and in what direction is the S.P.A.R.R.O.W. now moving (relative to its original motion)? Is he significantly off course? Explain your answer.

1. Prepare

If you're working your way straight through the sample problems, I highly suggest that you take a moment to stand up, stretch, and walk around a little bit. Give your legs some exercise, and get your blood flowing before you start in on another problem!

This one is a little bit more difficult, as we'll be dealing with angles and directions. In the previous problem, we got off easy. Everything was moving in the same direction. Here we'll have to do more work so we can account for the different directions stuff is moving. We can do it, however!

2. Draw two pictures

Okay, so here "the Birdman" is flying through space, rocking out to his mixtape of Aerosmith, Rush, and Lady Gaga[163]. Suddenly, the collision sensors go off – he's headed straight toward a giant asteroid! The computer calculates a 0.12 percent chance of survival... that's NOT good!

Let's draw a picture[164] of this!

163 Seriously, you need to hear the mashup "Born by Night Walking This Way."

164 Okay, a note of confession: I had entirely too much fun playing around with GIMP while drawing this picture; that's why it's a bit more flashy than the other diagrams. Have fun with your physics. It won't kill you... much.

As an experienced pilot, Jonathan knows exactly what to do! He fires his starboard thrusters, releasing a fast-moving stream of gas. The gas moves directly rightward — relative to his current path – and pushes Jonathan's S.P.A.R.R.O.W. to the left enough to miss the asteroid. Whew! Time for another picture[165]:

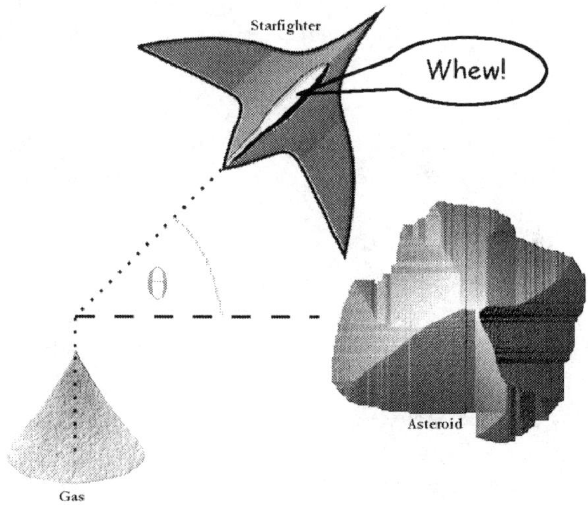

Jonathan can go back to enjoying the music... right?

165 For those in the know, the ship shouldn't really be rotated (strictly speaking); consider that a bit of artistic license!

3. Identify the momenta

At this point, we can already see that this problem is going to be more complex than the previous one. The S.P.A.R.R.O.W.'s path after releasing the gas isn't along any neat lines. In fact, we don't even know what direction the S.P.A.R.R.O.W. is moving, which is one of the things the problem is asking us to figure out. However, we can still write down general descriptions for how the objects are moving; this will help us later as we're building our equations.

Before the collision, the S.P.A.R.R.O.W. is moving forward (to the right in our picture). The reaction mass (the gas) is sitting quietly in a tank on board, waiting patiently to be released. In other words, both the starfighter and the gas are moving together and in the same direction. (That's actually a clue for the next step...) We'll call the momentum of the ship and gas before the collision $p_{together}$.

After the collision, the gas is moving straight rightward[166], as the problem says "at a right angle to [the ship's] original motion." The momentum of the gas will be p_{gas}. The starfighter, on the other hand, is moving forward and to the left. We just don't know how much of that motion is forward and how much of it is leftward[167]. How do we know that the S.P.A.R.R.O.W. has to be moving forward and to the left? Well, there's no reason for the ship to stop moving forward. (Remember Newton's first law?) If the ship doesn't come to a stop, then the release of gas to the right suggests that the ship should move in the opposite direction, to the left. (Remember Newton's third law?) If we put these together, we'll find that immediately after the collision, the spaceship should still be moving forward, but it should also be moving sideways and to the left. In any case, we'll name the starfighter's final momentum as p_{ship}.

4. Determine the elasticity of the collision

As I pointed out in the last step, the gas starts out on board the S.P.A.R.R.O.W. and moves with it. Because they start out together, then

166 In the picture, this direction is down.

167 "Forward and to the left" correspond in the picture to "rightward and toward the top of the page" respectively.

separate during the "collision" – which in this case is really an ejection – this collision is automatically considered perfectly inelastic. Because of this, we won't have to worry about the kinetic energy in the next few steps.

5. Choose a system of coordinates

For this problem, our choice of coordinate system is slightly more important than it was in the last problem. After all, we now have three directions to consider: the direction of the starfighter's momentum before releasing the gas, the direction of the starfighter's momentum after releasing the gas, and the momentum of the gas after release. Because these are not all at right angles to each other, we're going to have to break at least one of these into components.

That said, two of the momenta are perpendicular to each other: the original momentum of the starfighter and the momentum of the gas after the collision. If we choose those two directions to form our coordinate system, we only have one momentum to break into components! Let's update the picture with everything we have so far:

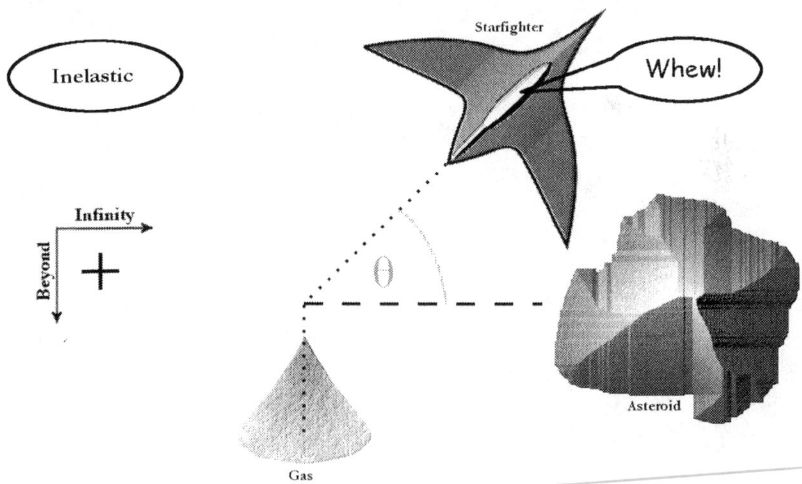

Again, I'm only showing the "After" picture to save space. From here on out, we'll use this coordinate system for everything we do.

6. Break momenta into components

In the first example, we didn't have any momenta that didn't lie along one of our coordinate directions. Now we need to work out the components for the momentum of the ship after the collision, as it's flying off at some unknown angle.

You might be concerned that we don't know what that angle is. The numerical value isn't so important. Instead, we need to focus on making sure that the angle has a variable – a name to hang onto. While I was tempted to name this angle *Weave* (as the partner to *Bob* from the Forces section), I instead decided to go with a more traditional θ. In fact, I've already added it to the picture. Don't worry if your drawing[168] didn't have the angle labeled yet; you can add it in now. (And please feel free to name it whatever you like!) Time to use SOHCAHTOA!

Based upon where I've put θ, the "Infinity" component of p_{ship} is the adjacent side of the triangle, while the "Beyond" component will be the opposite side of the triangle. Using SOHCAHTOA to break up the ship's momentum, let's find the coordinate components of all of the momenta.

	Before:	*After:*
"Infinity":	$+\mathbf{p_{together}}$	$+\mathbf{p_{ship}\cos(\theta)}$
"Beyond":	*0, nada, nothing, zip.*	$+\mathbf{p_{gas}}$, $-\mathbf{p_{ship}\sin(\theta)}$

Note that the "Beyond" component of the ship's momentum is negative. We defined positive "Beyond" to be to the right (or downward in the diagram). Since the S.P.A.R.R.O.W. is moving left (or upward in the diagram) its momentum is automatically negative.

168 You have been playing along at home, right? This is not a spectator sport! </scolding>

I've skipped a few steps to get to this point, so please make sure that you can reproduce this set of momenta! Hopefully it's clear how I got here.

7. Identify initial conditions

Now we're ready to add up our momenta by direction. If this collision had been elastic, we'd be considering the kinetic energy as well. Because it's not, however, we won't. So there!

"Infinity": $\mathbf{p_{together} = m_{together}\, v_{together}}$

"Beyond": 0, nothing, nada, zip

Wow, that was quick! Let's move on to the final conditions.

8. Identify final conditions

Just like in the last step, we just need to read the information from the table on the previous page.

"Infinity": $\mathbf{+p_{ship}\cos(\theta) = m_{ship}\, v_{ship}\cos(\theta)}$

"Beyond": $\mathbf{+p_{gas} - p_{ship}\sin(\theta) = +m_{gas}\, v_{gas} - m_{ship}\, v_{ship}\sin(\theta)}$

Short and sweet is quite the treat. Let's finish the physics off!

9. Apply conservation of momentum (and...)

Momentum is conserved by direction, which means that the initial total momentum in a given direction must equal the final total momentum in that direction. We'll get two equations then – one for each direction:

"Infinity": $m_{together} v_{together} = m_{ship} v_{ship} \cos(\theta)$

"Beyond": $0 = m_{gas} v_{gas} - m_{ship} v_{ship} \sin(\theta)$

Note: We have two equations and two unknowns, so we should be good to go! At this point, all we have left to do is to figure out the speed of the ship, v_{ship}, and the direction it's moving, θ!

10. Do math and solve

This is where things get tricky. Solving for angles in algebraic problems can be ridiculously annoying. I've found that the best way is to first eliminate everything but the angle. Once you've done that, you might be able to simply plug it in and find all the other variables. Just to keep it simple, let's go with the most straightforward method. I want to eliminate v_{ship}, so I'll solve for that first from the "Infinity" equation:

$$v_{ship}=\frac{m_{together}v_{together}}{m_{ship}\cos(\theta)}$$ [169]

I can then plug this into the "Beyond" equation to get

$$0=m_{gas}v_{gas}-\frac{m_{together}v_{together}\sin(\theta)}{\cos(\theta)}$$

The sine of any angle divided by the cosine of the same angle is the tangent of that angle. In other words,

$$0=m_{gas}v_{gas}-m_{together}v_{together}\tan(\theta)$$

A little bit more algebraic massage gives us the angle

$$\tan(\theta)=\frac{m_{gas}v_{gas}}{m_{together}v_{together}}$$

Normally I'd encourage you to stick with symbols until you're ready for your final calculations. However, in this case I'm going to make an excep-

169 Believe it or not, this equation is actually the seventh gem!

tion. Plug in the values now and find the angle θ. That way, you're not dragging inverse tangents all over the place!

To do this, we need to know how much gas was released. Given the rate of release (5.75 kg/s), and the time over which it was released (eight seconds) we can figure out the total mass of gas (46 kg). This points out an interesting ambiguity. Does the mass given for the starship include the gas's mass? Or is the mass of the fast gas separate, lass? If the mass value given in the problem statement does not include the gas, that value is m_{ship}, and I'll have to add in m_{gas} to find $m_{together}$. If it does include the gas, however, that value is actually $m_{together}$. In this case, to find m_{ship} I'll need to subtract out the mass of the gas. This would normally be a great time to ask for clarification. But because we have a little bit of time, we can run through both and see how they differ.

If the problem statement gives m_{ship}, the angle of the ship's motion must be

$$\theta = 8.57°.$$

Using the first equation on the previous page gives us a final value for the ship's speed of

$$v_{ship} = 1440 \text{ m/s}.$$

If, however, the problem statement gives $m_{together}$, the angle is instead

$$\theta = 8.60°.$$

Using the first equation on the previous page gives us a final value for the ship's speed of

$$v_{ship} = 1441 \text{ m/s}.$$

Wow, not much difference! Note, however, that figuring out the discrepancy didn't require us to completely redo the problem. Instead, we simply changed what we plugged into the variables to account for the different possibilities. The physics is the same, but the calculated values came out slightly different, based upon the assumptions we have to make to solve the problem.

Finally, the question asks you to determine whether "the Birdman" is significantly off course and provide an explanation for your answer. This is a question of interpretation and justification. In other words, choose an answer (yes, no, or maybe so), then justify it based upon the results you've found and your understanding of physics.

As an example, I could easily see justification for a "no" answer. After all, in both cases the angle was small – less than 10°. His speed also hasn't significantly changed – he's moving less than two percent faster than he was before. On the other hand, if I consider Newton's First Law, I could also see justification for a "yes" answer. After all, he's not moving in the same direction as he was before. If he goes far enough, he might end up quite a ways from where he had intended (remember that he'll continue going in a straight line unless and until an outside force changes his motion). It's kind of like throwing a ball to a friend. If he's right next to you, it doesn't matter so much if your aim is off a little bit. If he's far away, however, even a small error will put the ball out of his reach. Given the above, I'd probably argue a "maybe" answer. It really depends on the circumstances and how much longer he has to travel.

Booyah! Two done, and one left to go!

Example 3

Ormond Harriott is playing marbles with her granddaughter, Dasia. Dasia fires her marble (m = 3.18 g) at 3.77 cm/s, and it collides head on with Ormond's stationary marble (m = 4.25 g). (a) How fast is Ormond's marble moving just after the collision? (b) How fast is Dasia's marble moving just after the collision?

1. Prepare

This is another great place to get up and stretch. Not only does it give your butt a much-needed break, but a little bit of exercise can give your brain a kick as well. I'm not suggesting you go jogging or anything, but a good stretch and short walk will do wonders, especially after working on physics for a long time. Refresh your snacks and drink, too!

2. Draw two pictures

This one should be relatively straightforward, right? Two marbles enter, one marble leaves! Oh wait, that's the Thunderdome[170]. Let's draw a picture of the marbles instead. Before the collision, Dasia's marble is headed straight for Ormond's stationary marble, like this:

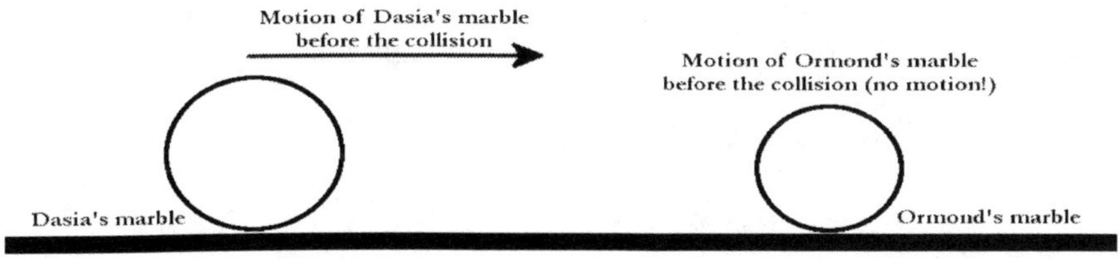

After the two marbles hit each other, Ormond's marble will definitely start moving and, not surprisingly, in the same direction Dasia's marble was going. After all, if someone "rear-ended" you, you'd move forward from the push. However, what will Dasia's marble do? There are three possibilities. Dasia's marble could continue moving the same direction, but slower. It

170 You know, playing marbles would be much more entertaining if it involved bungie cords and chainsaws...

could also end up going the opposite direction, if it hit Ormond's marble then bounced off. Or perhaps Dasia's marble could even come to a complete stop, transferring all of its momentum and energy to Ormond's marble!

Without practice and experience, it can be difficult to know for certain which will happen. Thankfully, though, we don't have to know., we only need to guess. Because it's easy, I'm going to choose to say that Dasia's marble will continue in the same direction. If I'm wrong, it should become clear after I work through the problem[171]. The after picture (which includes my guess) is here:

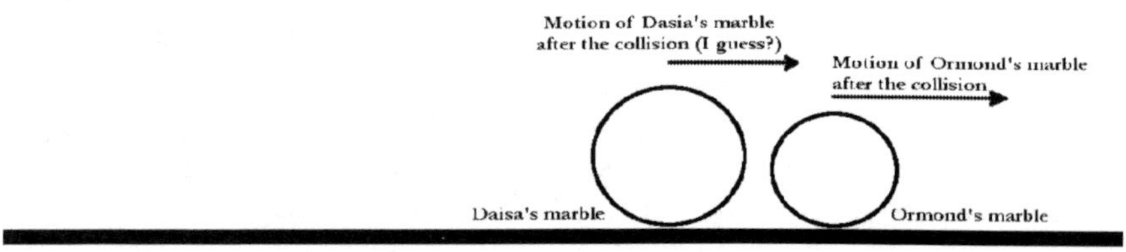

3. Identify the momenta

Based upon my two images (and my guess), Dasia's marble initially has momentum to the right and Ormond's marble has zero momentum because it's not moving. After the collision, both marbles have momentum to the right. Note, however, that the momentum of the two marbles are separate both before and after the collision. At no point can we treat them as one object. That will come into play in the next step.

4. Determine the elasticity of the collision

We're at the aforementioned next step, which is now the current step! Because the marbles are separate before the collision and are still separate afterward, this collision must be either partially inelastic or perfectly (practically) elastic. While there are a few clues that can help distinguish which approach is necessary, I'm going to give you a more straightforward method: Assume the problem is perfectly (practically) inelastic. Doing so will require more work (you will have to keep track of the kinetic energies be-

171 This is actually a feature; when I finish my answer will be correct regardless of whether or not my guess was correct!

fore and after the collision), but if you don't need them it will become obvious during the last step. Either you'll be able to solve the problem without using the kinetic energy equation, or you'll need to include some form of external energy somewhere. In either case, you'll have everything you'll need, and you won't have to backtrack and redo your previous work.

Following my own advice, the collision described in this problem must therefore be perfectly (practically) elastic.

5. Choose a system of coordinates

Since the momenta of the marbles is (basically) to the right all through this problem, I'm going to choose that direction as one of my coordinate axes. That way I won't have to break up any of the momenta into components. I'll call that direction "Classic," and the perpendicular coordinate axis, "Diet" (since I won't have anything to do with it). Here is the picture of my shiny new coordinate system:

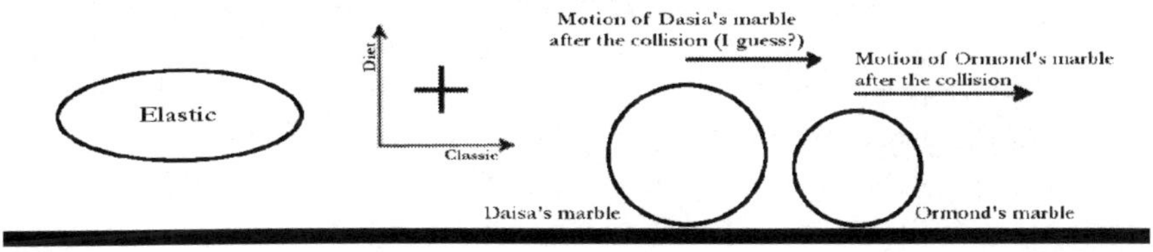

You may be wondering why I insist on putting in a second coordinate axis when there's obviously no momenta in that direction. In a nutshell, it's a habit. Each step of the conceptual scaffold is important, even if the degree of importance varies from problem to problem. By making sure that I always add and label two coordinate axes at right angles to each other, I reinforce the habit of performing the steps correctly. In other words, by doing a little bit of extra stuff on the problems that *don't* need it, I make it easier to remember to do that same stuff on the harder problems that *do* need it. Success in physics, then, comes not from a single act, but a series of applied habits. Okay, enough philosophy! Daylight's burning, and this physics problem isn't going to solve itself[172].

172 Although, strictly speaking, the evolution of reality over time could be considered a "self-solving physics problem"... Hmm, I'll have to think on that.

6. Break momenta into components

Because I based my coordinate system on the momentum specific to my problem, this step will be easy. Everything points in the positive "Classic" direction:

	Before:	*After:*
"Classic":	$+\mathbf{p}_{\textbf{dasia-initial}}$	$+\mathbf{p}_{\textbf{dasia-final}}, +\mathbf{p}_{\textbf{ormond-final}}$
"Diet":	*0, nada, nothing, zip.*	*0, nada, nothing, zip.*

There's not much more to say, so let's move on.

7. Identify initial conditions

So far, this problem has looked a lot like the first one. This step is where we'll finally have to take into account the fact that the collision is perfectly (practically) elastic instead of perfectly inelastic. While energy is always conserved, in perfectly (partially) elastic collisions, the kinetic energy is more specifically conserved[173]. Therefore, to solve the problem, not only will we need to determine the initial total momenta by direction, we also need to figure out how the kinetic energy is distributed.

We can pull the momentum information straight from the table we built in the last step:

"Classic": $\mathbf{p}_{\textbf{dasia-initial}} = \mathbf{m}_{\textbf{dasia's}}\,\mathbf{v}_{\textbf{dasia-initial}}$

"Diet": *0, nothing, nada, zip*

Now we only need to worry about the kinetic energy before the collision. Kinetic energy is energy of motion; the only object moving before the

173 If you were wondering about my choice of names for the coordinate system, think about "Conservation of Kinetic Energy," and hopefully the answer will "pop" to mind. Also, please realize that we're really just using Conservation of Energy, and all of the energy terms are kinetic energies.

collision is Dasia's marble. So the initial total kinetic energy is directly associated with that motion:

KE: $\frac{1}{2}\, m_{dasia's}\, v^2_{dasia\text{-}initia}$

If this turns out instead to be a partially inelastic problem, we might or might not need this. However, it's better to have it ready just in case.

8. Identify final conditions

After the collision, both marbles are moving, so both marbles should be represented in our final total kinetic energy. As before, we can get the momenta from the table we built in Step 6. I'll just add the kinetic energy stuff at the end:

"Classic": $\mathbf{p_{dasia\text{-}final} + p_{ormond\text{-}final}}$

$$\mathbf{= m_{dasia's}\, v_{dasia\text{-}final} + m_{ormond's}\, v_{ormond\text{-}final}}$$

"Diet": 0, nothing, nada, zip

KE: $\frac{1}{2}\, m_{dasia's}\, v^2_{dasia\text{-}final} + \frac{1}{2}\, m_{ormond's}\, v^2_{ormond\text{-}final}$

Again, in a partially inelastic problem, we may or may not need this part. If we do, you can be assured we'll have to add pieces here to account for the energy that was lost from kinetic during the collision.

9. Apply conservation of momentum (and...)

We're finally ready to apply our conservation laws! Unlike the last two problems, however, we not only have to conserve momentum in each direction, we also have to account for the kinetic energies. I'll drop the "Diet" direction because nothing's going on there. With the other two directions we find

$$m_{dasia's} v_{dasia-initial} = m_{dasia's} v_{dasia-final} + m_{ormond's} v_{ormond-final}$$

$$\frac{1}{2} m_{dasia's} v^2_{dasia-initial} = \frac{1}{2} m_{dasia's} v^2_{dasia-final} + \frac{1}{2} m_{ormond's} v^2_{ormond-final}$$

10. Do math and solve

At this point, we have two equations and two unknowns: $v_{dasia\text{-}final}$ and $v_{ormond\text{-}final}$. Like the last problem, we may have a little bit of difficulty with this. However, this time it's because of the squares in the second equation instead of annoying trig functions.

Let's start by solving for $v_{dasia\text{-}final}$. I'm not proposing this because I have a particular plan in mind, but because we have to start somewhere.

$$v_{dasia-final}=\frac{m_{dasia's}v_{dasia-initial}-m_{ormond's}v_{ormond-final}}{m_{dasia's}}$$

Now we can plug this into the second equation. Normally I'd give the simplified result, but here I'm going to show some of my steps to make sure we remain on the same page. Here's what we'll start with:

$$\tfrac{1}{2}m_{dasia's}v_{dasia-initial}^2=\tfrac{1}{2}m_{dasia's}\left(\frac{m_{dasia's}v_{dasia-initial}-m_{ormond's}v_{ormond-final}}{m_{dasia's}}\right)^2$$
$$+\tfrac{1}{2}m_{ormond's}v_{ormond-final}^2$$

Whew, the equation was so long I had to drop part of it to the next line. Next, I'll expand the portion in parentheses, multiply both sides by $m_{dasia's}$ (to get rid of the denominator), combine like terms, and then factorize. A lot is happening here, so please make sure you can get from the equation above to this one below:

$$0=m_{ormond's}v_{ormond-final}(m_{ormond's}v_{ormond-final}-2m_{dasia's}v_{dasia-initial}$$
$$+m_{dasia's}v_{ormond-final})$$

Note that there are only two ways for this equation to consistently be equal to zero. The first is if

$$v_{ormond-final}=0$$

If that's true, then it doesn't matter what's in the parentheses. (We already know that the mass of Ormond's marble isn't zero.) However, that would physically imply that Ormond's marble wasn't moving at all after the collision. That's kind of weird, right?

Well, if you think back to our discussion on possible types of collisions, I mentioned that there was another type of interaction: "not interacting at all." This is the possibility we've found here. If somehow Dasia's marble passed through Ormond's, both marbles would continue with the exact same motion they had before the collision. Ormond's marble would continue to sit there, just as this solution predicts. If you factor this zero into the solution for the speed of Dasia's marble, you'll find that the final speed is equal to the initial speed – Dasia's marble continues merrily on its way.

While this is perfectly fine and physically realizable, it's not the situation described in the problem statement. Dasia's marble hit Ormond's marble head on! So, based upon the interpretation of the result, we have to throw this out and instead consider the alternate solution. In that case, everything within the parentheses must somehow always be equal to zero:

$$0 = m_{ormond's} v_{ormond-final} - 2 m_{dasia's} v_{dasia-initial} + m_{dasia's} v_{ormond-final}$$

Here we can solve for the final speed of Ormond's marble:

$$v_{ormond-final} = \frac{2 m_{dasia's}}{m_{dasia's} + m_{ormond's}} v_{dasia-initial}$$

Great! We're almost done. Plugging this back into the equation for the final speed of Dasia's marble we find:

$$v_{dasia-final} = \frac{m_{dasia's} - m_{ormond's}}{m_{dasia's} + m_{ormond's}} v_{dasia-initial}$$

Aha! This actually answers a question we had back in Step 2, which was "what direction will Dasia's marble be moving after the collision?" Now we have our answer. According to this equation (which we found using our understanding of physics), if the mass of Dasia's marble is larger than that of Ormond's marble, then the final speed of Dasia's marble will be positive. That means that it will continue going in the same direction! If, however, the mass of Dasia's marble is less than that of Ormond's marble, the final speed will be negative, which indicates that Dasia's marble will have reversed directions. We can quickly check the marble masses in the problem statement – Dasia's marble is 3.18 g whereas Ormond's is 4.25 g; Dasia's marble is doing the moonwalk!

In addition, an interesting case occurs when the two marbles are equally sized. In that case, Dasia's marble would come to a complete and utter stop! Because it would have lost all of its motion, it would have also lost all of both its kinetic energy and its momentum. The only thing that could have received them (due to conservation of energy and conservation of momentum) would have been Ormond's marble, so after the collision, Ormond's marble would have been moving with the speed Dasia's marble had before the collision. This is actually the physics behind a Newton's cradle – you know, that desk toy that has five metal balls hanging in a row. If you let one swing, when it hits those at the bottom it stops, and the one on the far end goes flying off. If you let two swing, then two go flying off, and so on. It gets really freaky when you raise all but one of them and then let go. (If you've never tried it, you should!)

Anyway, let's go ahead and finish this problem by plugging everything in:

a.) $v_{ormond\text{-}final} = 3.23$ cm/s to the right

and

b.) $v_{dasia\text{-}final} = 0.543$ cm/s to the left.

Note that it's perfectly valid to replace the negative sign with "to the left"; remember, the sign gives us information about the direction of the motion relative to the coordinate axis – positive signs are in the same direction as the axis and negative signs are in the opposite direction.

Before I make my usual triumphal statement about finishing another solution, I'd actually like to point something out. Strictly speaking, this solution is wrong.

Oh Em Gee! What?!

Yes, I know. You're horribly disappointed. I am, too. But let me at least explain what I mean! Because this problem mentions marbles, it could be reasonably assumed that the marbles are rolling. Rolling requires that the marbles turn, which is a form of rotational motion. Linear motion consists of moving sideways, forward, backwards, etc. in straight lines (hence the "linear"). Rotational motion involves going around in circles. As a type of motion, however, it has kinetic energy and momentum associated with it! Which means that when we were doing the work of figuring out the initial and final momenta and kinetic energies, we should have included terms (and possibly whole equations) to account for the rotational piece of the motion! I didn't do that. Instead, I solved the problem as if the marbles were sliding across the floor instead of rolling.

There are a few reasons why this isn't as bad as it sounds. First, the point of the example problems is to practice the conceptual scaffolds. In that respect, this example is good. Second, although rotation does add some complexity, the conceptual scaffolds still work as written. No major changes result from adding rotational motion.

So why didn't I go into it? Well, the physics of rotation requires a lot of exploration, especially once you consider the different possible ways something can rotate. Is it rotating about its center or about some point on the surface? Is it rolling along a surface, or is it spinning in the air – "free space," if you will? How many directions is it spinning in, and what does it mean to spin in a given direction? Answering all of these questions sufficiently would require several chapters – in effect, an additional section, just on rotational motion. I considered it strongly for a long time, and even wrote an extensive outline including that material. Eventually, I came to the conclusion that the book would be stronger without it. Anyone who can master the three conceptual scaffolds included in this book has the basic framework necessary to solve any physics problem – including rotational motion problems.

Afterword

Congratulations! By completing this book you've started mastering tools that will bring you amazing results – not just in solving physics problems, but in many areas of your life. Remember, the conceptual scaffolds are just a starting point – something to imitate as you begin to learn how to approach and systematically reduce problems into small, more manageable pieces. As you begin to assimilate these ideas into your problem-solving toolbox, you'll find yourself improvising more and more. Perhaps one day you'll get a problem that looks like a mashup of a forces problem with a conservation of energy problem, and you'll have the tools to do it! Just remember to take your time and break it into those manageable parts. I've said it before, and I'll say it again: You *can* do it! Eventually you'll be able to innovate and create your own conceptual scaffolds to solve whatever problems you regularly face.

As you work your way to that point, here are a few final suggestions that will help you survive introductory physics:

1. Don't rush! Take enough time to fully consider the problem. The first and obvious answer is very rarely correct.

2. Practice makes perfect! Take the time to work through a variety of problems. Take chances and make mistakes. As you see different problems and different solutions, you'll start recognizing patterns in the way some ideas fit together. Then, if you rush (and ignore my previous advice), you'll be more likely to get it right, based upon the intuition you've built through experience.

3. No, seriously, don't rush if you don't have to. Nothing replaces being careful and thorough.

So what happens once you've survived your first semester physics course? Well, for some of you, that's it! You're ready to move on to other pastures of intellect and inquiry, and I wish you all the best of luck.

For those of you who are moving on to the second semester physics course, however, don't panic. You now have a tool that is more useful than a Babel fish and more portable than a hitchhiking guide: practice in applying scientific thinking! Now that you have experience working with systematic reductionist problem-solving approaches (i.e., the conceptual scaffolds), you've got a leg up on any future physics class and, I'll venture to say, any science class that you take in the future. Second semester physics focuses on electromagnetic phenomenon, but really it just introduces new forces, new energies, and new objects to interact through conservation of momentum. In other words, it's the exact same basic processes with a fresh coat of pain.

Err, that really should be paint. Seriously.

Anyway, even if you stick with physics, and end up studying waves and optics, quantum mechanics, statistical mechanics, or even magnetohydrodynamics, you'll still be mostly dealing with forces, energies, and momenta. The conceptual scaffolds in this book will still be of great use to you.

And if you ever run into a situation that doesn't yield to one of those, you will still be okay, because you'll be able to use the skills you've learned using the conceptual scaffolds to improvise – you'll know what works, what doesn't, and be able to build your own scaffolds to help you solve the problem. You'll be okay. I have great faith in you.

Finally, I just want you to remember. The force will be with you. Always. And so will the energy and the momentum, but who's counting, right?

T.H. Cooke

Appendices

Appendix: Kinematic Equations

If you happen to be one of those folks who is interested in the actual derivation of the kinematic equations, you've come to the right place. Hopefully you've done some work on calculus lately, because we're going to be using a lot of it! Let's jump right in... ready? GO!

We'll start with the definition of acceleration: change in speed over some-time interval.

$$a = \frac{\Delta v}{\Delta t}$$

Unfortunately, this only tells us about changes over a long time frame. It doesn't tell us how the speed is changing every single instant. However, we can figure that out if we shrink the time interval to zero:

$$a = \lim_{\Delta t \to 0} \frac{\Delta v}{\Delta t}$$

If you remember your calculus, this is the same thing as the derivative, or

$$a = \frac{dv}{dt}$$

Now that we have a differential equation, we can effectively integrate to find the speed as a function of time:

$$v(t)=\int a\,dt$$

If the acceleration changes, we're done, because without knowing how it behaves over time, we can't do the integration. For our purposes here, we'll assume that the acceleration is constant. With that assumption, we can do the integral to obtain the speed equation:

$$v(t)-v(t_0)=a(t-t_0)$$

which is usually written as

$$v(t)=v_0+a\Delta t$$

From here, we can apply the definition of speed as a change in position over some time interval:

$$v=\frac{\Delta x}{\Delta t}$$

Doing the exact same trick with limits as we did before also produces a derivative:

$$v=\frac{dx}{dt}$$

which we can integrate to find the position as a function of time:

$$x(t)=\int v(t)\,dt$$

Like before, we can't actually calculate the integral unless we know how the speed changes over time. Fortunately, we found that earlier when we were working with the acceleration. Note, however, that this restricts us to those earlier limitations as well – the equation we find will only hold if the acceleration is constant. Doing the math, we find the equation for the position as a function of time:

$$x(t)=x_0+v_0\Delta t+½a\Delta t^2$$

That gives us two of the three! The third one just requires a little bit of algebra. So far, we've found how the velocity and position change as a function of the time interval. The last kinematic equation shows how the speed changes as a function of position. We're going to combine these first two equations, eliminate the time interval, and do some algebra to produce the third kinematic equation.

We'll start by solving for time in the speed equation:

$$\Delta t=\frac{v(t)-v_0}{a}$$

We can then plug that into the position equation:

$$x(t)=x_0+v_0\left(\frac{v(t)-v_0}{a}\right)+½a\left(\frac{v(t)-v_0}{a}\right)^2$$

Multiplying everything out gives:

$$x(t)=x_0+\frac{2v_0v(t)-2v_0^2}{2a}+\frac{v^2(t)-2v_0v(t)+v_0^2}{2a}$$

(Note that I also went ahead and got a common denominator for the second and third terms – we'll need them to simplify this!) You may see some terms that are redundant, so let's clean things up some:

$$x(t)-x_0=\frac{v^2(t)-v_0^2}{2a}$$

The left side of the equals sign is the change in position, so a little more algebra gets us to:

$$v^2(t) = v_0^{\,2} + 2a\,\Delta x$$

Appendix: Calculus Example

Here you will find an example of a calculus-based problem and its solution using the force conceptual scaffold. As always, the problem statement is presented at the beginning so that you can try applying the conceptual scaffold by yourself, then check your work against the solution.

Example: The other shoe drops; unfortunately, it's from a great height. The 425-gram shoe presents an average cross-sectional area of 0.0175 m^2 as it falls. The constant of drag for air against the shoe is 7.35 Ns/m^3.

a. Assuming it was dropped from rest, how fast is the shoe going after four seconds?

b. What is the shoe's terminal velocity?

Use the following equation to calculate the magnitude of the force of air drag:

$$F_{drag} = c A_{shoe} v$$

Example

The other shoe drops; unfortunately, it's from a great height. The 425-gram shoe presents an average cross-sectional area of 0.0175 m^2 as it falls. The constant of drag for air against the shoe is 7.35 Ns/m^3.

a. Assuming it was dropped from rest, how fast is the shoe going after four seconds?

b. What is the shoe's terminal velocity?

Use the following equation to calculate the magnitude of the force of air drag:

$$F_{drag} = c\,A_{shoe}\,v$$

1. Prepare

Alright. Take a deep breath. Even though this is a calculus-based example, and we must use calculus to solve it, there is no need to panic. After all, it's still just physics, right?

Based upon the problem statement, I think it's safe to assume that we'll need to use the forces conceptual scaffold. If you haven't reviewed the algebra-based examples, I highly suggest you do so now.

2. Draw a picture

Based on my assumption (and fondest wish) that you've already reviewed several sample problems, I'm not going to reproduce the same image over and over again (that's not where the new and interesting stuff is!). Instead, I'm going to describe in steps how I got the final image on page 232.

Let's first determine the object of interest. In this case, it's the shoe. Now, as you may have noticed throughout this book, I'm not such a great artist; in addition, I'm feeling pretty lazy right now. So I'm going to go with the old physicist's standard picture: a circle. That's right, a circle. After all, why torture yourself drawing a bunch of details if they really don't matter much to the physics? For almost any problem that does not involve rotation, you can replace whatever object you're drawing with a circle, or a rectangle, or a square. Trust me, it won't make a difference to your calculations. And if you want to get fancy, you can even use a triangle or a rhombus.

3. Identify the forces

The problem statement just screams "gravitational force." After all, a shoe is falling from a great height! Therefore, weight is an obvious force to add to our drawing. In addition, drag is specifically mentioned, along with a helpful equation to use for calculating the size of the force! So we should most definitely add drag as well.

The problem doesn't explicitly state which direction the drag should point. However, that's not too hard to figure out. The drag force results from a falling object colliding with many, many air molecules as it falls. Although this is horrifically complicated at the molecular level, when we zoom out from the object, the cumulative effect is clear: The air resists the motion of the shoe, just like friction resists sliding motion. Therefore, the drag force will point in the exact opposite direction that the shoe is moving. Since the shoe is moving downward as it falls, the drag force must be upward.

At this point, you might be second-guessing my choice of a circle for the picture. After all, shoes are kind of lopsided and bulbous; won't that have a huge effect?

Technically that is correct. In addition, as the shoe falls, the uneven forces on it from the air will cause it to tumble wildly. However, the problem statement gives us the average cross-sectional area of the shoe. In other words, although it's tumbling and turning, the problem expects us to gloss over those details by looking at the average motion of the shoe. So our circle idea still works... at least for a basic solution.

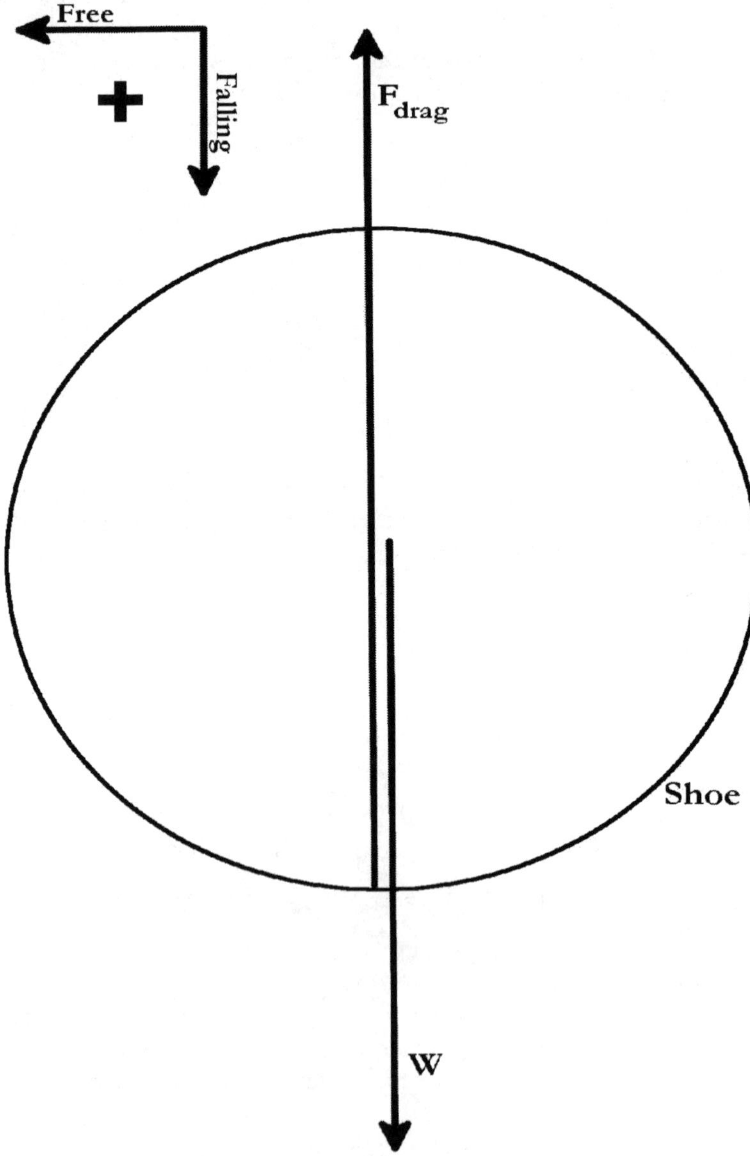
Free
Falling
+
F_{drag}
Shoe
W

4. Choose a coordinate system

The shoe is falling downward, so let's choose that as one of our special directions and call it "Falling." Because nothing is happening in any other direction, we could stop there; however, I like to choose at least two axes when setting up a coordinate system, just in case. So I'll also set up a "Free" coordinate that's perpendicular to "Falling." After all, the shoe is free. It's free falling[174].

5. Break forces into components

Because the drag force and the weight point along the coordinate axes, we don't have to worry about breaking the forces into components for this problem. That said, be aware that in a more complex situation – say, modeling the motion of a Wiffle ball shot horizontally out of a cannon – you may need to break up the forces into components at the start of the problem *and* break up the object's motion (velocity) into components. Because the magnitude and direction of the drag force relies on the direction and speed that the object is moving, you have to be very careful to properly track all of the vector quantities.

However, we'll stick with our simpler problem. The drag force is in the negative "Falling" direction. The weight is in the positive "Falling" direction. Whew!

6. Apply Newton's second law

$$\vec{a} = \frac{\sum \vec{F}_{ext}}{m}$$

The great thing about physics is that regardless of whether you're doing calculus-based or algebra-based physics, the core ideas (and equation statements of those ideas) remain the same. So even though we're dealing with drag, we'll still rely on our old buddy, Newton. The big difference will be in what comes out once we apply his law.

174 And I offer my most sincere apologies to Tom Petty.

Combining everything we've learned, Newton's second law gives us

$$a_{falling} = \frac{+W - F_{drag}}{m}$$

That doesn't look so bad; it's comparable to what we might find in an algebra-based problem. But what happens when we start using math?

7. Do math and solve

At this point, we're ready to start using the mathematical relationships we've been provided to solve the problem. Let's start by plugging in the equations for the weight and the drag force:

$$a_{falling} = \frac{mg - c\,A_{shoe}\,v}{m} = g - \frac{c\,A_{shoe}}{m}v$$

Houston, we have a problem. Acceleration is defined as how an object's motion changes based upon external forces. Here, however, the acceleration is partially determined by the motion itself. In other words, the motion determines how the motion is changing! It's like the movie Inception, only in physics! Fortunately, we've got this covered. By putting in the definition of acceleration as the change in speed over time, and rearranging the equation we can get

$$\frac{dv}{dt} + \frac{c\,A_{shoe}}{m}v - g = 0$$

This is a first order differential equation. There are a lot of ways to solve for v. My favorite has to be what I call "the wild-ass guess". The way this works is – brace yourself – we guess at a solution based upon mathematical experience. Then we plug it in to see what's necessary to make it work.

Here, for example, I'm going to guess that the speed, as a function of time, looks something like:

$$v = B + C e^{-Dt}$$

B, C, and D are constants, and will hopefully be determined by the limitations of the problem as stated. From this, the acceleration should be

$$a = \frac{dv}{dt} = -D C e^{-Dt}$$

If I plug these values into the equation I got from Newton's second law, I finally get something I can work with:

$$\left(-D C e^{-Dt}\right) + \frac{c A_{shoe}}{m}\left(B + C e^{-Dt}\right) - g = 0$$

with some simplification I end up with these two separate terms:

$$\left(\frac{c A_{shoe}}{m} - D\right) C e^{-Dt} + \left(\frac{c A_{shoe}}{m} B - g\right) = 0$$

Note that the right side of the equation does not depend on time. In other words, if our understanding of physics is correct – meaning we've properly used the conceptual scaffold and correctly applied Newton's second law– then this equation is, by definition, true, regardless of what t is. There is only one way for this to happen: Both of the quantities in the parentheses must be equal to zero. Then t doesn't matter because it will always be multiplied by a zero! In other words,

$$\frac{c A_{shoe}}{m} - D = 0 \quad \text{and} \quad \frac{c A_{shoe}}{m} B - g = 0$$

Remember how I said those constants should be determined by the limitations of the problem as stated? We just found those limitations:

$$D = \frac{c A_{shoe}}{m} \quad \text{and} \quad B = \frac{m g}{c A_{shoe}}$$

Plugging this back into the original guess for the speed, we see that

$$v = \frac{m g}{c A_{shoe}} + C e^{-\left(\frac{c A_{shoe}}{m}\right) t}$$

At this point we're almost done! The only piece of the puzzle we don't have is C, but regardless of what it is, we know that this equation for the speed adequately satisfies Newton's second law. How do we figure out what that last constant is?

Well, there's one piece of information that we haven't used yet – the shoe started from rest. That means that when t was equal to zero, v was also equal to zero:

$$0 = \frac{m g}{c A_{shoe}} + C$$

A little bit of algebra allows us to solve for C:

$$C = -\frac{m g}{c A_{shoe}}$$

Plugging that back in, we have our final equation for the shoe's speed,

$$v = \frac{m g}{c A_{shoe}} \left(1 - e^{-\left(\frac{c A_{shoe}}{m}\right) t}\right)$$

Answering part a) is now very easy. To calculate the speed of the shoe after four seconds we just plug and chug to get v = 22.7 m/s.

Part b) is more difficult, but not by much. The **terminal velocity** of the shoe is determined by how fast the shoe is moving after a long time of falling. In other words, what does v look like when t is big? Well, the exponential of a big number is about zero, so we can entirely drop that term out of the equation. That gives us an equation for the terminal velocity as

$$v_{terminal} = \frac{m\, g}{c\, A_{shoe}}$$

Actually plugging in the values gives a terminal velocity for this shoe of 32.4 m/s., or 72.5 mph. For comparison, the terminal velocity of a tumbling human being is about 120 mph.

Finally, just to clear the record, falling doesn't hurt. It's the landing that gets you. In other words, if you're worried about surviving a fall you shouldn't worry about the forces that caused you to fall. Instead, worry about the forces that will cause you to *stop* falling. Anything that reduces those forces will help you out!

Appendix: Philosophical Musings and Geometry Reminders

Fair warning: Everything I state in this appendix is my personal opinion and point of view. There are others who would disagree with these ideas.

Mathematics is a language, just like French, Spanish, or English. It has a set of structural rules (a "grammar," if you will), and a defined set of meanings (a lexicon). There are two big differences between math and French, however. First, the grammar of math doesn't just define the structure of meaningful statements. It also defines a set of logical transformations that can be performed upon those statements. For example, addition, subtraction, multiplication, division, and exponentiation are all part of the grammar of math. We may call it algebra, or calculus, or group theory. However, these are all shorthand ways of describing the set of rules that we're using – the grammar of math!

Similarly, there's a dictionary of mathematics. For example, "+" represents the logical transformation of addition. There's one symbol, however, that is in some ways key to reading and understanding math: the "equals" sign, "=." Assuming I was writing a dictionary of math, the entry for the equals sign would look like this:

= (equals): (verb)

1. is the same as (?)
2. is equivalent to
3. is defined as

The first definition is what most people think of when they see "=." What's on the right side of the equals sign is the same as what's on the left side. However, this may or may not be true! Like English statements, math statements may be false. For example, although the statement

Red is blue

is a grammatically correct sentence, it is not a true statement. Similarly,

$$6 = 4$$

is structured correctly, but violates the meaning of "6" and "4." The grammar is right, but the meaning is wrong. So the first form of "=" can be thought of as a question, kind of like "is equal to?" In fact, in some situations, math teachers encourage writing a question mark over the equals sign to make this uncertainty explicit. This form of the equals sign is evaluative.

The second definition is more of a command: This thing on one side is a direct consequence of that thing on the other side. This is the "equals" of calculation – when I calculate this thing, that thing is what I get. This statement of equals shows a causal relationship: Based upon the definitions, these two things must be the same. In other words, it implies compliance with the grammar of mathematics. This is the causative form, and the statements it involves are always true as long as the grammar of mathematics is correctly followed. (In other words, if a causative equals sign forms a false statement, it's your fault.)

Finally, the third version is an assertion: a statement of strict definition. There may or may not be any logic behind the assertion, depending on who's writing it. For example, let's say I have the equation

$$x = y^2 + 3y - 2$$

There's no particular reason that x should equal that quadratic equation in y. Yet, with that equals sign I am asserting, by personal fiat, that it does. This form of the equals sign is therefore assertive. Because the equals sign forms the "verb" of the mathematical statement, it's important to understand what it might be saying.

Before we get into reading sentences, there's one more thing I want to point out about the mathematical lexicon. Like other languages, math has pronouns – substitutes for a different, more clearly defined object. In math, they're called variables.

Most languages, however, have a maximum of about four base pronouns: "he", "she", "it", and "they". ("They" might come in "he" and "she" versions, but then there's usually no "it.") Mathematics, however, has an undefined number of these – in fact, math allows you to define a new one whenever you want! This leads to both the power and the frustration of using mathematics. To understand the equations, you need to understand the definition of the pronouns (variables) that appear in the equations. If you don't understand what they're referring to, then it's impossible to gain any understanding or meaning from the statement! For example, what does the following sentence mean?

She told her that she took it to her, but she said that she hadn't given it even though she said that she had taken it then.

If you know what I'm talking about there, please let me know what's going on, because even trying to write that sentence confused me greatly!

In the same way, it's really important to know what the variables of a mathematical equation refer to so that you can make sense of it.

Math as the Language of Physics

Based upon this understanding of mathematics, it should be obvious why physicists like to speak in that language. Being able to define your own pronouns allows an incredible amount of both precision and flexibility in making statements about the real world. In addition, we already tend to use measurements of the physical world based upon the lexicon of mathematics. If we combine that with the power of the grammar of math, we can make predictive statements based upon physical ideas.

Let me give you an example. Consider Newton's first law:

The acceleration of an object is directly proportional to the total external force applied to that object, and inversely proportional to the object's mass.

This is a statement of physical interaction. Now, let's translate that statement into mathematics by creating a dictionary of pronouns:

a represents acceleration

F represents force

m represents mass

Then, using the assertive equals sign, we can use math to make the same statement as the bold quote above:

$$a = \frac{\sum F_{external}}{m}$$

I haven't included directionality, but it wouldn't change much. Mathematics has built-in symbols to show that a given pronoun (variable) has both direction and magnitude, and the grammar includes rules to handle that. (Breaking vectors into components is one way of tackling it.)

At its core, then, the whole point of the conceptual scaffolds is to help you take a statement of physics, translate it into math, then apply the grammatical rules of mathematics until you've gotten the shortest possible statements of causative equals. In other words, you examine the physics, write your equations based upon it, then do algebra until you've solved for your unknown variables. Ideally, there should be a last step of translating that final math statement back into English. For introductory physics, however, it's not usually necessary. If you can successfully do the initial translation and the simplification, then you're right on track!

Geometric Relationships to Remember

If you're having trouble with the finding angles in the "Break <something> into components" steps, look over the following true statements. In general, these will cover all of the geometry you will need to solve introductory physics problems[175]:

Statements of Definition

- Right angles are 90°. Always. No, seriously – *always.*
- Parallel lines do not cross. *Ever.*
- Perpendicular lines always cross at right angles. *I mean it.*
- The interior angles of triangles always add up to 180°. *For great justice!*

For Intersecting Lines[176]

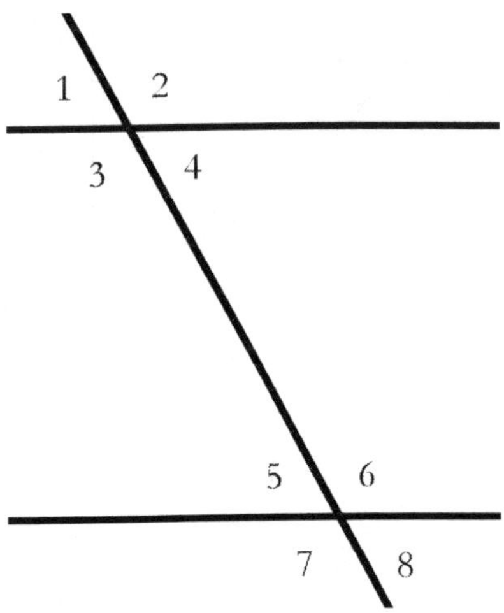

- Angle 1 and Angle 2 add up to 180°, as do Angle 1 and Angle 3. These are called **supplementary angles**.
- Angle 1 and Angle 4 are the same, as are Angle 2 and Angle 3. These are called **opposite angles**.
- Angle 4 and Angle 5 are the same, as are Angle 3 and Angle 6. These are called **alternate interior angles**.

175 Note: these rules assume a Euclidian (flat) geometry. If you're in some sort of complex curved space, all bets are off, although most of these will still work at small enough scales.

176 In the diagram, the two horizontal lines are parallel.

Index

G

I

L